Early Adventures in Biochemistry

FOUNDATIONS OF MODERN BIOCHEMISTRY
A Multi-Volume Treatise, Volume 1

Editors: **MARGERY G. ORD AND LLOYD A. STOCKEN,** *Department of Biochemistry, University of Oxford, Oxford, England*

Early Adventures In Biochemistry

By: **MARGERY G. ORD**
LLOYD A. STOCKEN
Department of Biochemistry
University of Oxford
Oxford, England

JAI PRESS INC.

Greenwich, Connecticut *London, England*

Library of Congress Cataloging-in-Publication Data

Foundations of modern biochemistry / editors, Margery G. Ord and Lloyd A. Stocken.
p. cm.
Includes bibliographical references and indexes.
ISBN 1-55938-960-5 (v.1)
1. Biochemistry—History. I. Ord, Margery G. II. Stocken, Lloyd A.
QD415.F68 1995
574.19'2'09—dc20

95-17048
CIP

55 Old Post Road, No 2
Greenwich, Connecticut 06836

JAI PRESS LTD.
The Courtyard
28 High Street
Hampton Hill
Middlesex TW12 1PD
England

ISBN: 1-55938-960-5

Library of Congress Catalog Number: 95-17048

Transferred to Digital Printing 2006

CONTENTS

ACKNOWLEDGMENTS

We are very grateful to our colleagues in the Department of Biochemistry who have encouraged us in our attempt to present in this volume the history of biochemistry before 1960. We are especially grateful to Professors Ed Southern and George Radda for allowing us space in the department to continue working. Brian Taylor, the departmental librarian, and the staff of the Radcliffe Science Library have been very helpful and patient in obtaining books for us and checking references. Ken Johnson helped us with the illustrations and photography, and the advisory staff of the Computing Service and Dr. J. Sanders courteously and efficiently extricated us from word-processing crises.

Drs. Mary Gale, Brian Lloyd, and the late Hugh Sinclair helped us with references to early studies on nutrition; Mr. Reg Hems and Dr. Dereck Williamson recounted their memories of working with Sir Hans Krebs; and Dr. F. L. Holmes very kindly allowed us to read in manuscript the first volume of his authoritative work on Krebs. The late Professors Bill Paton and David Whitteridge directed us to important references in the history of physiology. Professor Bradford, the

Archivist of the Biochemical Society, Dr. P.J. Fitzgerald, Professor Joel Mandelstam, and Dr. Michael Yudkin kindly read various drafts and made valuable suggestions. We are also particularly grateful to Dr. Michael Foster and Dr. Bruce Henning for their care in reading and correcting the manuscript. Any mistakes are ours, but they hopefully have been minimized thanks to the assistance of all of our friends in reviewing the material.

The photographs of early researchers in biochemistry are reproduced with the kind permission of: Elsevier Science Publishers; The Nobel Foundation; The Department of Biochemistry, University of Oxford; Oxford University Press; Annual Reviews, Inc.; L.A. Stocken; and Pergamon Press.

Margery G. Ord
Lloyd A. Stocken
Editors

Chapter 1

INTRODUCTION

This book is intended for students of biochemistry, biology, and medicine who are familiar with textbook knowledge of intermediary metabolism. Present-day graduates however, are often unaware of the contributions made to this knowledge by the great biochemists in the earlier part of this century (see also Kennedy, 1992). We hope this volume will help to correct this deficiency and strengthen interests in these pioneers. We have also tried to show how our present information about some of the central pathways in animals was obtained, describing the limited experimental techniques which were available and indicating how advances in methodology opened up new areas of the subject which were then enthusiastically explored. The account covers the period from 1900 to 1960, but also outlines the principal developments in earlier centuries from which biochemistry emerged (Chapter 2). We have not attempted a rigid historical treatment; the findings are considered in the light of our present knowledge. For convenience, current flowsheets for the pathways are included (Appendix 2).

In Chapter 2 we shall see how biochemistry developed from the interactions of organic chemistry and physiology—the study of organized living systems—and the consequent attempts to explain the behavior of cells in chemical terms. By the 1880s the recognition by Koch and Pasteur of the roles of bacteria in infectious diseases, and the growth of immunology following increasingly widespread and successful vaccination against smallpox, led to the foundation of specialized institutes for medical research. The Koch Institute for Infectious Diseases opened in 1880, the Pasteur Institute in Paris in 1882, the Russian Institute for

Experimental Medicine in St. Petersburg in 1890, and the Kitasato Institute in Tokyo in 1892-3, commemorating von Behring's discovery, with Kitsato, of diphtheria antitoxin. In New York, the Rockefeller Institute for Medical Research was incorporated in 1901 and in London the Lister Institute for Medical Research was established by 1903.

Support for biochemical investigations came also from the demands of industry. In Germany, organic chemists analyzing fermentation were involved in, or were consultants to, the brewing industry. Pasteur in the 1850s was employed to advise French viniculturists. Both in the U.K. and the U.S. requirements of agriculture prompted studies in animal nutrition; Rothamsted Experimental Station was set up in 1843.

The different laboratories were often visited by those who became the pioneers of biochemistry in their home countries. An even greater cross-fertilization came in the 1930s in the U.K. and the U.S. from the arrival of biochemists from Europe. Their experience of techniques in intermediary metabolism laid the foundations for work in this area in their host countries in the 1940s and 1950s.

Systematic investigation of biochemical events within organisms began with respiratory and calorimetric measurements on small animals and man, leading to the determination of the animal's energy requirements and the identification (ca. 1900-1940) of factors which were essential in the diet, notably vitamins (Chapter 3). Clinical studies on patients with hormone disorders as in juvenile diabetes (insulin insufficiency), Addison's disease (adrenal cortical insufficiency), goitre, and exopthalmia (thyroid disturbances) and examination of changes in patients' blood and urine (the most readily available human materials), gave some insight into the effects which endocrine organs had on human metabolism (Chapter 3).

By the start of this century physiologists, who were almost invariably medically qualified, were investigating the behavior of isolated organs such as heart or gastrocnemius muscle, especially the contractions caused by electrical stimulation and responses to drugs. These experiments were greatly facilitated by the availability of suitable salt solutions (1880s, Ringer) in which the organs could be maintained for considerable periods without apparent deterioration in their properties. It was not however easy to extend these studies to an intracellular level.

With the exception of Buchner's yeast extract and some comparable muscle preparations (Chapter 4), disrupting tissues often caused such damage to cells that normal metabolism was irreversibly affected. A further obstacle was that classical methods of analysis were neither sufficiently sensitive, rapid nor simple enough for the multiple measurements required to follow chemical changes in small samples of tissue.

Introduction of photoelectric cells led to the replacement of the Duboscq colorimeter and so to quantitative spectrophotometric methods of analysis which met biochemical requirements. This introduction of spectrophotometry as a routine procedure was one of the earliest technological advances underpinning the elucidation of biochemical pathways between 1930-1960. Micromanometric methods also became available about the same time, and offered a means to measure cell respiration. Manometry was developed in Warburg's laboratory in Berlin and was one of the main techniques used by H.A. Krebs in his studies on the citric acid and urea cycles (Chapters 5 and 6).

Because disrupted tissue preparations were unsatisfactory, attempts were made to work either with more organized systems such as tissue slices (liver-Krebs) or to identify and isolate the intracellular organelles involved in the reactions. Cytochemical procedures were developed in the 1930s and 1940s to locate sites of reaction *in situ* in cells (Chapter 9). Examination of cell ultrastructure became possible when the electron microscope was introduced after 1945. Techniques for the isolation of cell organelles, notably mitochondria, were developed about this time (Chapter 9).

Isotopes were introduced into biochemistry by Hevesy in the 1920s. Their use was essential for the studies of Schoenheimer and Rittenberg which showed that most of the cell constituents were in a dynamic state, constantly being broken down and resynthesized (Schoenheimer, *The Dynamic State of Body Constituents*, 1941). Using isotopes, investigations of pathways of biosynthesis for carbohydrates and lipids followed rapidly, the carbon cycle in higher plants being formulated by Calvin in the 1950s. As scintillation counting was introduced for the measurement of weak emitters, particularly ^{3}H and ^{14}C, establishment of pathways became a routine procedure (Chapter 8). The study of protein synthesis and of the role of nucleic acids followed inevitably (Watson, 1968). Once nucleic acids could be sequenced, analysis of the blueprints which specify and select cell constituents became possible and molecular genetics was born—the start of another story.

As metabolic pathways became clearer, the detailed study of the enzymes involved was facilitated by the introduction of new procedures for isolation, purification, and characterization of proteins. Developments in chromatography in the early 1940s and the introduction of gel electrophoresis allowed more efficient methods to be used to separate proteins and to analyze their primary structure, so that Sanger was able, by 1953, to report the primary structure of insulin (Chapter 10).

Classical biochemistry is in the main the study of cytoplasmic enzymes and the interacting pathways of reactions which they catalyze. Until the mid-1940s those engaged in these studies had been trained primarily as chemists or physiologists. Their numbers were small, personal contact was commonplace, and there were only a few biochemical journals in which to publish the observations. Courses in physiological chemistry were available in two thirds of the medical schools in the U.K. by 1909, but biochemistry was only available as an undergraduate science course in a few universities, and many universities did not have separate departments for the subject. Only after World War II did biochemistry become an undergraduate subject in its own right. Its popularity was stimulated in part by the impact made by the introduction of sulphonamides and especially penicillin and from the hope derived, for example, from the study of vitamins, that diseases might soon become explicable on a molecular basis and thus curable. A further influence, particularly for physical scientists, came from Schrodinger's book, *What is Life?* (1944), which prompted some physicists and chemists to turn their minds to biological problems. For those already engaged in the subject, Ernest Baldwin's *Dynamic Aspects of Biochemistry* (1st Edition, 1947), offered an exciting attempt to bring the newly discovered pathways into a coherent, integrated scheme.

Towards the end of the twentieth century, interest in biochemistry has shifted to molecular genetics and its widespread applications. The study of metabolic pathways has become a relatively small part of the subject.

While these reactions form an essential element in modern biochemistry and cell biology, the way they were established experimentally is seldom described. A knowledge of the limitations of the procedures available at the time when the reactions were discovered and of the conceptual contexts into which the new data had to be accommodated, may help biochemists to appreciate how their subject began and the radical changes there have been in the latter half of the twentieth century.

BIBLIOGRAPHY

In addition to the references after each Chapter, many of the articles in early editions of Annual Reviews of Biochemistry, Advances in Enzymology, Advances in Protein Chemistry, International Reviews of Cytology, Physiological Reviews, Vitamins & Hormones, and other review serials, refer to specific topics considered in the text. Most of the articles we have cited give an overview of the topics. Where these are available many individual references have been omitted.

General

Autobiographical memoirs of distinguished biochemists in the first chapter of each Annual Review of Biochemistry.

Biographical Memoirs of Fellows of the Royal Society.

Boyer, P.D., Lardy, H., & Myrbach, K., Eds.(1963). The Enzymes, 2nd ed. Academic Press, New York.

Florkin, M. & Stotz, E.H., Eds. (1962). Comprehensive Biochemistry, Vol I. Elsevier, Amsterdam.

Gabriel, M.L. & Fogel,S., Eds.(1955). Great Experiments in Biology. Prentice-Hall, Englewood Cliffs, NJ.

Greenberg, D.M., Ed. (1954). Chemical Pathways of Metabolism. Academic Press, New York.

Kennedy, E.P. (1992). “Sailing to Byzantium.” Annu. Rev. Biochem.61,1-28.

Liebecq, C., Ed. (1956). Conferences et Rapports, 3rd Congres International de Biochimie.

McElroy,W.D.& Glass, B., Eds.(1951) Phosphorus Metabolism. Johns Hopkins Press, Baltimore.

McElroy,W.D.& Glass, B., Eds.(1954) Mechanisms of Enzyme Action. Johns Hopkins Press, Baltimore.

Summer, J.B.& Myrbach, K., Eds. (1951) The Enzymes. 1st ed. Academic Press, New York.

Walker, M.D., trans. (1888). Dr.Schussler's Biochemic Treatment of Disease (Boericke, W.& Dewey, W., Eds.). 2nd ed. Billing & Sons, Guildford, UK.

Watson, J.D. (1968). The Double Helix. Atheneum, London.

Chapter 2

BIOCHEMISTRY BEFORE 1900

THE ANALYTICAL BASIS OF BIOCHEMISTRY

The word, "biochemie," was coined by Hoppe-Seyler in his introduction to the first volume of *Zeitschrift fur Physiologische Chemie* in 1877. It was used similarly by Dr. Schussler in *Biochemic Treatment of Disease* (1888) where biochemistry was defined as "the science dealing with the chemical actions, or the chemistry in the tissues of living things." It might now be defined as the study of the chemical and physicochemical properties and processes of living cells and their constituents. Not only was the term "biochemistry" unknown before the last quarter of the nineteenth century, but the definition implies a physicochemical, mechanistic interpretation of living phenomena which, before 1900, was not generally accepted, and indeed was vehemently opposed. Chinese, Indian, Greek, and later, Arabic cultures were concerned with notions such as "lifeforces," "breath of life," "humors" (Needham, 1970; Leicester, 1974). The emergence of biochemistry as we now understand it depended on the growth of modern chemistry from the latter half of the seventeenth century (see Teich, 1992).

It is possible to identify several apparently disconnected themes which led ultimately to the analysis of the chemistry of living cells. Perhaps the first biochemical experiment in Europe may have been that of van Helmont (1579-1644), a physician with a profound interest in chemistry. He grew a willow tree in a weighed amount of earth, watered it for five years and then weighed the earth and the tree again. The weight of the earth was unchanged but the willow had increased in weight by 163 pounds, from which van Helmont concluded water was the element out of which everything was created (Leicester, 1974).

In the mid-seventeenth century independent studies by Boyle, Lower, and Mayow indicated atmospheric air was involved in the same manner in combustion and respiration and that breathing air generated animal heat. One hundred and fifty years earlier, Leonardo da Vinci had compared animal nutrition to the burning of a candle, observing that animals could not survive in an atmosphere that would not support combustion. Boyle deduced air supplied a substance necessary for life. Lower showed that something in the air was responsible for the brighter color of arterial than venous blood. Mayow recounted in his book, *Respiration* (1669), that a mouse breathing in a closed chamber used up some part of the air. If a flame was burning in the chamber the mouse died almost immediately. Mayow also believed that alterations in the composition of air were effected in the lungs.

It was not until 1774 that experiments on the burning of charcoal led Priestley to realize that the changes in the air described by Mayow were due to the exhaustion of what was later to be called oxygen. The product of the combustion of carbon with oxygen was called "fixed air." Lavoisier in 1777 correctly explained Priestley's and others' experiments on combustion, showing that when mercury was burned, it gained as much weight as the air lost. Scheele's "fiery air" and Priestley's "dephlogisticated air" were the identical essential component of air, oxygen, which was utilized in respiration or combustion to yield "fixed air" whose composition Lavoisier established as carbon dioxide. In 1780 he and Laplace then carried out the first recorded metabolic experiment. A guinea pig was placed in a closed container surrounded by ice. Oxygen consumption and the amount of melted ice were measured and a direct relation found between the amount of heat produced and the oxygen consumed. Further experiments by Lavoisier showed that many substances of plant or animal origin could be combusted in the presence of air or oxygen to yield carbon dioxide and water. He concluded that respiration was a reflection of oxidation—"La vie est une fonction chimique."

Table 1. Some Natural Products Isolated Before 1830

Date	*Compound*	*Source*	*Investigation*
	Ethanol	Wine	
	Acetic acid	Vinegar	
	Glycerol	Animal fat	
1770	Tartaric acid	Grapes	Scheele
1775	Benzoic acid	Benzoin	Scheele
1776	Uric acid	Bladder stones	Scheele
1780	Lactic acid	Milk	Scheele
1783	Oxalic acid	Wood sorrel	Scheele
1784	Citric acid	Lemons	Scheele
1786	Malic acid	Apples	Scheele
1773	Urea	Urine	Roelle
	Cholesterol	Gallstones	Poulleltier de Lasalle
1805	Morphine	Opium	Serturner
1810	Cystine	Urinary calculi	Wollaston
1814	Casein	Milk	Berzelius
1820	Glycine	Gelatine	Braconnet
1820	Quinine	Plants	Pelletier & Caventou
	Strychnine		
	Brucine		
	Cinchonine		

A second extremely important strand was the development (ca.1800-1830) of reproducible techniques for the quantitative elementary analysis of natural products. By 1810 Gay-Lussac and Thenard had shown that sugar, gum, starch, milk-sugar, oak, and beechwood contained only carbon, hydrogen, and oxygen, with hydrogen and oxygen being present in the proportions found in water (see Lowry, 1936). Compounds such as sugar and starch became classified as carbohydrates. Some of the earliest natural products to be isolated, mainly from plant sources, are shown in Table 1.

Attempts to adapt Lavoisier's methods to obtain reproducible analyses of carbon, hydrogen, and oxygen contents were frequently unsuccessful until improvements were introduced by Liebig in 1831 in the combustion process. The other major constituent in natural products, nitrogen, had been identified by Cavendish (1779) as the main component in the gas remaining after the combustion of charcoal in air. Quantitative determination of nitrogen was introduced by Dumas (1830). Some 50 years later, Kjeldahl introduced an alternative procedure for the determination of nitrogen titrimetrically which, on a microscale, became a standard technique to estimate the nitrogen content of proteins and other cell constituents, and was used with only slight modifications at least until the 1950s. By the end of the nineteenth century most of the main classes of natural products had been identified. When the elementary composition of a substance had been determined, its properties were explored and its structure normally confirmed by synthesis.

The great impact of the quantitative analytical approach on the development of structural organic chemistry made Liebig one of the most influential chemists of his time.

Since then organic chemistry has become a cornerstone of biochemistry.

VITALISM AND THE CELL THEORY

By the early years of the nineteenth century various properties of living systems had been described: movement of organisms, contractility in muscles, excitability (irritability) of nerves, sensitivity, and secretion, as well as respiration in animals, fermentation in yeast, and photosynthesis. These phenomena stopped at death, or when the structure of the organ(ism) was disrupted. Inorganic chemistry was associated with inanimate matter and was sharply distinguished from the carbon-based chemistry characteristic of living organisms. Stahl (1660-1734) who as a chemist advanced the phlogiston theory, believed that all the features which distinguished living from dead bodies were conferred by anima, an immortal principle which after death returned from whence it came.

By the end of the eighteenth century the importance of organization had been recognized by physiologists. The Ecole de Sante at Montpellier, especially Bordeu (1722-76) and Barthez (1734-1806), believed that a vital principle was the basis of all life, which was inherently associated with organization and was "the totality of forces opposed to death" (Bichat, 1802). While vitalism could easily be accommodated with religious beliefs, Barthez in particular distinguished between the soul and the vital principle. "When a man dies his body goes back to [its] elements, his vital principle is reunited to that of the universe and his soul goes back to God, who gave it to him ..."

In contrast to the organizational hypothesis of some physiologists, Liebig believed the vital force was a physical force controlling the formation of living systems by opposing chemical forces which after death led to decomposition and putrefaction. The constituents of living organisms were thought by Liebig to be held together by weak forces. When no longer protected by the vital force these constituents, in the presence of oxygen in the air, underwent molecular movements so breaking the weak interactive forces.

Interpretation of the process of fermentation by yeast was one of the most controversial issues for vitalists. Its resolution was fundamental for the future development of biochemistry. In the early nineteenth century fermentation was believed to be related to putrefaction and decay. Liebig considered it to result from the breakdown of a substance (sugar) following the admission of air to the nitrogenous components in yeast juices. After the must of grape juice had fermented, the liquid cleared and the yellow sediment, yeast, was deposited.

The start of what would prove to be conclusive evidence against Liebig's views on fermentation came from microscopical observations made possible by improvements in instrument design early in the nineteenth century (see Chapter 9). By 1824 Dutrochet had proposed "All organic tissues are actually globular cells of exceeding smallness, which appear to be united only by simple adhesive forces; thus all tissues, all animal (and plant) organs, are actually only a cellular tissue variously modified." The presence of the nucleus as the essential characteristic of plant cells was recognized by Brown (1833). Following further work by Schleiden and especially by his friend

Schwann, the cell theory emerged. ". . cells are organisms and entire animals and plants are aggregates of these organisms arranged according to definite laws" (Schwann, 1838).

In his studies on fermentation, Schwann (1837) suggested putrefaction was due to "living germs" in the air. He showed that if a yeast suspension was heated, and heated air passed through it, there was no fermentation. Schwann recognized the living nature of yeast and linked fermentation to multiplication of the organism. The following year another microscopist, Latour, also reported that fermentation depended on the presence of yeast, as did Kutzing, whose paper was unfortunately lost by the editor, although his observations were subsequently recovered from his notebooks.

By 1839 Schwann was distinguishing between the combination of molecules to form cells and those phenomena which resulted from chemical changes either in the component particles of the cell itself or in its surroundings.

> These may be called metabolic phenomena. [Metabolism] is an attribute of the cells themselves [with] vinous fermentation an instance of this. [Further,] each cell is not capable of producing chemical changes in every organic substance... but only in particular ones. The metabolic power of cells is arrested not only by powerful chemical action, [which] destroys organic substances in general, but also by matters which are chemically less uncongenial, [e.g.] concentrated solutions of neutral salts [or by] other substances in less quantity [e.g.] arsenic.

Schwann was also vigorously anti-vitalist, being unable to accept the idea of a force whose properties changed with the organ under study, exhibiting contractility in muscle, excitability in nerves, etc.

> We must ascribe to all cells an independent vitality, that is, such a combination of molecules as occur in any single cell is capable of setting free the power by which [the cell] is enabled to take up fresh molecules... The cause of nutrition and growth resides not in the organism as a whole but in the separate parts, the cells.

In 1838 a paper by Turpin confirming the observations of Schwann and Latour on yeast, was published in *Annalen der Pharmacie*, a journal edited by Liebig, Dumas, and Graham. Liebig was strongly opposed to the cell theory and to the postulated role of yeast in fermentation. He therefore arranged for a spoof, satirical paper to be published immediately following that from Turpin, caricaturing the interpretations of the cell theorists. This ridicule greatly upset Schwann, and prevented his appointment to professorships in Prussia. He therefore accepted a somewhat uncongenial appointment to the Chair of Anatomy in Louvain, and did little more outstanding work on cells.

Failure to accept the cell theory and the central role of cells in metabolism caused great difficulties when dynamic properties of cells were being investigated. An analytical approach which totally ignored the heterogenous cellularity of tissues, led to uninterpretable data on the composition of organs and their variations in disease. Attempts were made to extend quantitative methods to the study of chemical changes in organisms. Analyses of whole tissues such as heart were compared with those of metabolic end-products which were considered to be urea and uric acid in urine, and carbon compounds in bile. Liebig believed combustion of fat and carbohydrate (respiration) occurred in the blood and thought proteins were the only nutrients to be assimilated by animals. His influence was such that the adoption of Schwann's ideas on the role of cells was seriously delayed.

THE ACCEPTANCE OF THE CELL THEORY AND THE DOWNFALL OF VITALISM, 1850-1897

By the second half of the nineteenth century German chemists had established a dominant position in analytical and synthetic organic chemistry. Various simple sugars and aminoacids were being isolated and characterized, as well as more complex plant products. Studies on the composition of blood and the properties of hemoglobin were also well under way. The composition of lipid-rich components and the order of the different units within complex macromolecules, such as proteins and nucleic acids, could not however be resolved by techniques then available.

Laws of physical chemistry were also emerging; Helmholtz' paper *On the Conservation of Energy*, was presented in 1847. It and Hess'

Law of Constant Heat Summation (1840-1843) were to provide the theoretical foundations for the metabolic balance studies of Voit and his pupils (Chapter 3) which established the quantitative link between food consumption and energy output, and thus the beginnings of the modern study of nutrition.

The impact of the second law of thermodynamics (Gibbs, 1878; Helmholtz, 1882) on the interpretation of the structure and behavior of cells was less immediate. It was not until the publication of Lewis and Randall's text on thermodynamics in 1923 that concepts of free energy, enthalpy, and entropy became familiar to succeeding generations of biochemists, provoking in the 1940s and 1950s vigorous arguments between physical chemists and more pragmatic biochemists over the concept of "high-energy phosphate" (see Chapters 4 and 5).

For biologists and physiologists, improvements in the design of microscopes allowed more reliable interpretations of tissue structure. People were becoming familiar with the appearance under the microscope of sections of plant and animal material. Virchow, the founder of histopathology, strongly supported Schwann's earlier views of the cell as the ultimate unit of life. "What made Schwann immortal... is the demonstration of the cellular origin of all tissues" (Virchow, 1877). Virchow and von Kollicker (1852) recognized that visible and functional differences among tissues were due to their different cell types.

The most important contribution in this period to the acceptance of the cell theory and the correct interpretation of fermentation as a process caused by living organisms, was the work of Pasteur. In 1857 he was commissioned by French viniculturists to investigate the presence of lactic acid in their wine. Pasteur showed conclusively that the acid was produced by living cells or "ferments," which were distinct from those producing ethanol. Different microbial species caused different chemical reactions. His observations were in direct conflict with those of Berzelius who thought fermentation was due to contact with non-living catalysts (Chapter 10). Further, Pasteur realized that the decay of dead cells or the addition of nitrogenous matter was not the cause of the fermentation, as was maintained by Liebig, but served only as food for the growing cells. If the conditions of the fermentation were altered, the nature of the product changed, an acid reaction favoring the formation of ethanol and a neutral reaction favoring lactic acid.

Accompanying the acceptance of the cell as the unit of life was the abandonment of the idea of spontaneous generation. Its supporters argued that under the conditions used by Pasteur and others, life was not possible if, to maintain sterile conditions, air had been excluded. Further, heating the medium might inactivate it so that life could not be sustained. In 1859 the Academie des Sciences offered a prize for conclusive evidence against spontaneous generation. Pasteur's experiments, using swan-necked flasks, confirmed Schwann's earlier observations and showed that preheated infusions of yeast remained sterile unless contaminated after cooling. If the neck of the flask was broken so that its contents became contaminated, preheated air supported life, thus demonstrating that heating the air did not destroy its capacity to support life.

His experiments on fermentation established Pasteur as the "father of microbiology" (Marjory Stephenson, 1930). He remained a vitalist in the Montpellier tradition all his life. Fermentation was a vital process caused by living organisms —vital ferments. In the 1840s Pasteur had separated the enantiomorphic crystals of sodium ammonium tartrate. He considered the ability to induce optical activity was an exclusive and significant feature of living as opposed to non-living systems "establishing perhaps the only well-marked line of demarcation between the chemistry of living and dead matter. "Life...is a function of the dissymmetry of the Universe." It would be nearly 100 years before asymmetric syntheses were achieved non-enzymically.

Wohler's preparation of urea from ammonium cyanate, which could in principle be derived totally from inorganic constituents, is cited as an early demonstration (1828) that living cells were not obligatorily required for the synthesis of natural products. "I can prepare urea without requiring a kidney or an animal—either man or dog." Three years after the death of Pasteur the finding by Hans and Edouard Buchner (1897) that fermentation still occured in a cell-free extract from yeast and so did not require the presence of organized cells, was virtually the final nail in the coffin for vitalism and an essential preliminary to the study of intermediary metabolism (Chapter 4).

FROM PHYSIOLOGICAL CHEMISTRY TO BIOCHEMISTRY

During the early years of the 19th century there were disputes about the relevance of chemistry to medicine. In a lecture at the medical school in Philadelphia in 1818, Charles Caldwell, a professor of natural history, argued: "Shall I be told that chemistry aids in the explication of any of the phenomena or laws of the living body, either in a healthy or a diseased state?...I feel myself compelled to deny the position. [Chemistry] has superadded corruption to what it found already sufficiently corrupted." Organovitalists insisted life could not be explained by chemistry which was thought to be applicable only to dead matter. Others thought life too complex to be explained chemically. Physiologists maintained that a metabolic pathway could only be studied in the organism itself.

The work of Pasteur enforced the view that living cells could perform identifiable, complex organic reactions. At the same time physiologists such as Heidenhain and Bernard showed that metabolic changes associated with different organs could be performed *in vitro,* e.g., by gastric or pancreatic secretions. It became feasible for physiologists to interpret their observations in chemical terms.

In 1872 Hoppe-Seyler was appointed to the Chair of Physiological Chemistry in Tubingen and in 1877 inaugurated the first periodical to be devoted to "chemische lebensvorgangen"—the chemistry of living matter—*Zeitschrift fur Physiologische Chemie.* In the UK, Michael Foster held the Chair of Physiology in Cambridge. In 1898 he invited Gowland Hopkins to go to Cambridge to stimulate teaching and research on the chemical side of physiology. Hopkins was appointed to the Chair of Biochemistry in Cambridge in 1914. W.D.Halliburton, who was initially in the Department of Physiology at University College, London, moved to King's College in 1890 and established the first research school of biochemistry in the U.K. "securing for biochemistry [in the U.K.] general recognition and respect." (Morgan, 1983).

In the U.S. the *Journal of Biological Chemistry* was founded in 1905. The *Biochemical Journal* (U.K.) was begun by Benjamin Moore, then in the Chair of Biochemistry at Liverpool, in 1906 and the British Biochemical Society was founded by J.A.Gardner and R.H.A.Plimmer in 1911.

REFERENCES

Florkin, M. (1972). A History of Biochemistry. Parts I & II, Vol.30 of Comprehensive Biochemistry, (Florkin, M. & Stotz, Eds.) E.H. Elsevier, Amsterdam.

Florkin, M. (1975). ibid Part III, Vol.31.

Franklin, K. J. (1949). A Short History of Physiology, 2nd ed. Staples Press, London.

Hall, T.S. (1969). Ideas of Life and Matter. Vols.1 & 2. Chicago Press.

Leicester, H.M. (1974). The Development of Biochemical Concepts from Ancient to Modern Times. Harvard UP.

Lowry, T.M. (1936). Historical Introduction to Chemistry. Macmillan, London.

Morgan, N. (1983). William Dobinson Halliburton, FRS, (1860-1931) Pioneer of British biochemistry? Notes and records of the Royal Society 38, 129-145.

Needham, J. (1970). The Chemistry of Life. Cambridge UP.

Partington, J.R. (1964). A History of Chemistry. Macmillan, London.

Rothshuh, K.E. (1973). History of Physiology. Trans. (Risse, G.B., Ed.)

R.E. Krieger, Huntingdon, N.Y.

Teich, M. with Needham, D.M.(1992). A Documentary History of Biochemistry, 1770-1940. Leicester UP.

Chapter 3

EARLY METABOLIC STUDIES:
ENERGY NEEDS AND THE COMPOSITION OF THE DIET

THE DETERMINATION OF ENERGY NEEDS

By the 1840s the work of physical chemists was leading to the formulation of ideas of energy conservation and the laws of combustion and heat summation. As early as 1798 Count Rumford had observed that heat was produced when horses were working. On a voyage to Java, Robert Mayer noted that in the tropics, because of the body's decreased need for heat production, metabolic activities were less intense. Venous blood was redder in the tropics than in Europe from which he concluded it contained more oxygen and so had supported less combustion. These observations were the foundations of his formulation of the first law of thermodynamics "No given matter is ever reduced to nothing and none arises out of nothing" (1842). Joule and Helmholtz reached similar conclusions, and by 1858 Mayer, in an unpublished paper, could write: "The very same relations obtain between the combustion process on the one hand and the production of heat and force on the other. In the living animal carbon and hydrogen are oxidized and heat and motive power produced in return."

Calorimetric measurements from the mid nineteenth century were used to determine the amount of energy released in combustion (oxidation) of foods. Liebig (1842) demonstrated that carbohydrate, fat, and protein were all oxidized in the body and that corrections had to be applied to the amount of energy obtained from proteins because of their incomplete oxidation to urea. Urea excretion provided a convenient measure of protein utilization.

Studies on the balance between energy input as food and output as heat were commenced by Boussingault (1839) in cows. He analyzed the carbon, nitrogen, and oxygen contents of the fodder, the milk, and the excreta. Similar studies were performed by Voit (1857) on nitrogen equilibrium in dogs. Over a 58-day period he measured the amount of meat eaten, its nitrogen content and the nitrogen contents of the urine and feces. The balance between nitrogen input and output was confirmed, the figures differing by only 0.3%, underlining the accuracy maintained in the study.

Work by Voit and his associates continued so that by 1900 standard values for heats of combustion of different foods had emerged (Table 1). Respiratory quotients (RQ) were also derived, associated with the utilization of the different foods. The RQ is the molar ratio of the amount of carbon dioxide produced in the oxidation of a substance to the amount of oxygen needed for that oxidation. For carbohydrate the RQ is 1:

$$C_6H_{12}O_6 + 6O_2 = 6CO_2 + 6H_2O$$

For fatty acids, the RQ is about 0.7:

$$C_{15}H_{31}CO_2H + 23O_2 = 16CO_2 + 16H_2O$$

In the "post-absorptive state" with the subject at rest not less than 12 h after the last meal, protein catabolism has been completed. An RQ

Table 1.

	kJ/g	*kcal/g*
Protein*	17	4
Fat	39	9.25
Carbohydrate	16.3	4
Ethanol	29.7	7.1

Notes: *Corrected for incomplete oxidation to urea.

measurement between 0.7 and 1.0 therefore indicated what mixture of fat and carbohydrate had been oxidized. If the subject fasted so that the main energy source was fat, an RQ of 0.7 was found. This approach formed the background for the determination of energy balances in man by indirect calorimetry. In the post-absorptive state the RQ indicated the unique mixture of fat and carbohydrate being oxidized. This produced a known amount of energy/volume of oxygen utilized. On average the RQ in the post-absorptive state was 0.82. Measuring the oxygen consumed was therefore sufficient to determine the heat production over the period. With the subject completely at rest this energy was called the basal metabolic rate (BMR). By the end of the nineteenth century Rubner and others had shown that the BMR was affected by the age, sex, and surface area of the subject. It was also altered in certain illnesses, particularly thyroid diseases (see below).

Direct calorimetric studies on humans were attempted by Pettenkofer and Voit (1866). A room was constructed large enough to hold a man, so that his expired air could be metered and its carbon dioxide and water contents determined. The fasting subject was weighed and his water consumption and urine production measured. By 1902 Benedict and Atwater in the U.S. had constructed a calorimeter in which the subject could rest or undertake standardized exercise, with heat production being measured directly in the jacketed walls. Respiratory analyses could also be performed. Calorie intakes were therefore compiled for men working or resting under a variety of circumstances. Once the energy needs for humans had been determined it was possible to consider how the energy should be provided and what, if any, were the essential constituents in the diet.

Defined sources of food were therefore required. By 1905-1906 diets consisting solely of purified protein, carbohydrate and fat were shown to be inadequate to sustain life. Lunin (1881) and Pekelharing (1905) established that "white mice fed a bread baked with casein, albumin, rice-flour, lard, and a mixture of all the salts which ought to be found in their food, with water to drink, starve to death. For the first few days all is well, the diet is eaten, and the animals look healthy. But they all get thinner, their appetite diminishes, and in four weeks all are dead. If however, instead of water they are given milk to drink, they are kept in good health, though the [additional] quantity of albumin, lactose, and fat which they assimilate in the milk is negligible compared with [that in] the bread which they eat. ... There is an unknown substance in milk which, even in very small quantities is of paramount importance to nutrition. ... Undoubtedly this substance not only occurs in milk but in all sorts of foodstuffs, both of vegetable and animal origin" (Pekelharing).

These and other experiments of his own caused Hopkins in 1906 to conclude "No animal can live on a mixture of pure protein, fat, and carbohydrate and even when the necessary inorganic material is carefully supplied, the animal still cannot flourish." It therefore became the practice to supplement standard diets with milk so that control animals continued to grow or remain in nitrogen balance. Desirably the diet was available *ad libitum* but because animals on a deficient diet often eat less, "pair-feeding" was introduced. Here, freely available food eaten by the experimental animals on day 1 was weighed and only that weight of food was given to the control animals on day 2. Animals offered restricted amounts of food usually ate it all, thus ensuring that the only difference between the control and experimental groups was the nature of the diet not its total energy input.

DIETARY REQUIREMENTS

Protein

The need for a source of nitrogen in the diet was established by Magendie (1816) who put dogs onto a diet of sugar and water. By the second week the dogs were weakening, and in the third week developed a severe ulceration of the eyes, an early description of xerophthalmia and a sign of vitamin A deficiency. The dogs were dead within a month. To determine how much nitrogen was required Voit (1881) analyzed diets of apparently healthy soldiers and found them to have an average intake of 118 g protein per day. Siven (1901) at the age of 31, and weighing 65 kg, kept himself in nitrogen balance eating 25-30 g protein per day with a total calorie input (protein, fat, and carbohydrate) of 43 kcals per kg body weight. R.H. Chittenden (1904) suffered from a rheumatic knee joint. He reduced the protein content of his food to 37-40 g/day, without altering the extent of his physical activity and checked from his urinary urea output that he was staying in nitrogen balance. At the start of his 6-month experiment he weighed 65.5 kg. After reducing his calorie and protein intake he lost 8 kg in weight, and considered his rheumatic condition had been improved.

People with high protein intakes were also studied. At the start of this century Inuits were largely carnivorous eating mainly seal meat. They showed no tendency to increased vascular or renal disease, and thus became a continuing source of interest to dieticians. In 1921 Stefanssen wrote *The Friendly Arctic* describing his 11-1/2 years in the Arctic circle, for 9 of which he subsisted on the Eskimo diet. During this time he was in splendid physical condition. Stefanssen says in his book. "There is probably no field of human thought in which sentiment and prejudice takes the place of sound judgement and logical thinking so completely as in dietetics." (!) One of the puzzles about Stefanssen's experiences is how he avoided getting scurvy. This led to considerable discussion between nutritionists. In a later work, *The Fat of the Land* (1956), Stefanssen reviewed the ascorbic acid content of his diet. He suggested "If one has considerable fresh meat in his diet every day, and does not overcook it, there will be enough of whatever prevents scurvy to do the preventing"!

The best type of protein for the diet was studied by growth and nitrogen balance experiments. For this purpose growth was considered to be the orderly increase of all components of the living body. Rodents were principally used, the roughly linear increase in rat weight from weaning (ca. 30 g) to about 100 g being convenient as a measure of growth. The "biological value" of the protein gave an indication of its relative effectiveness as sole nitrogen source in the diet (Thomas, 1909). Initially casein was chosen as standard (Hopkins, 1906-1907) because of its availability and the ease with which vitamins could be removed by alcohol extraction. Later work showed casein was comparatively deficient in cystine/methionine and tryptophan (see below), egg albumin being preferable (biological value of ovalbumin for growing rats 97, for adult humans, 91; casein 69 and 56). Plant proteins which are deficient in a number of essential amino acids (see below) are significantly worse (wheat gluten, growing rats 40, adult humans 42; peanut proteins 54 and 56). The currently recommended intake of "good quality" (biological value 100) protein for an adult is 0.7-0.8 g/kg/day (1986).

Amino Acids

That protein *per se* was not required in the diet but only its constituent amino acids, became apparent early this century. Growth and nitrogen balance experiments showed that enzymically hydrolyzed casein was as good a source of nitrogen as the original protein, but that if the casein had been hydrolyzed by acid, the mice died (Henderson and Dean). By 1907 Henriques had shown that acid hydrolysis destroyed tryptophan, which had therefore to be replaced in the hydrolysate. Abderhalden (1912) extended these experiments using a mixture of alanine, arginine, aspartate, cystine, glutamate, glycine, histidine, isoleucine, leucine, lysine, phenylalanine, proline, serine, tryptophan, tyrosine, and valine. Nitrogen balance in the dogs was maintained but only over a relatively short period as threonine and methionine, which were later found to be essential, had not yet been isolated.

By the 1930s many workers had shown that nutritionally inadequate proteins, such as zein from maize, could be effective as a source of nitrogen if supplemented by additional amino acids (for zein, tryptophan). Even if it contained all the essential amino acids, the amount of protein in the diet influenced the results. Osborne and Mendel found that if the diet contained 18% by weight casein, which is low in cystine, young rats grew, but if the amount of protein was diminished, added cystine was required to offset the relative deficiency of this amino acid. Later, after methionine had been discovered, it was shown to replace the need for cystine.

The complete identification of the amino acids which are essential in the diet is due to W.C. Rose (1938). His first attempts to replace casein with its constituents were unsuccessful because an essential amino acid component in the protein hydrolysate had been missed. After threonine had been isolated by him from casein and fibrin, and shown to be essential, Rose identified val, met, his, lys, phe, leu, ile, thr, and arg as the necessary amino acids in rat diets. Arg was only needed for young actively growing animals. Rose then performed similar nitrogen balance studies on human (student) volunteers. Histidine, which is synthesized by microflora in the intestine, is not normally essential for man. Non-essential amino acids in protein can be replaced by isonitrogenous diammonium citrate.

Although these experiments showed growth was possible using casein hydrolysate, Rose also demonstrated that when the amino acid mixture was used rather than the intact protein, additional calories had to be provided as fat plus carbohydrate, if nitrogen balance was to be maintained. It was later shown that the carbohydrate was needed to protect the free amino acids from oxidation in the intestinal epithelium in the course of absorption. Further, amino acids are poorly tolerated by mouth, causing vomiting and/or diarrhea. After World War II attempts to feed very emaciated prisoners in concentration camps with protein hydrolysates were unsuccessful. It was then recognized that osmotic effects from the amino acids were responsible for the unpleasant consequences.

NUTRITIONAL DEFICIENCY DISEASES AND THE DISCOVERY OF THE VITAMINS

From very early times several diseases had been attributed to dietary inadequacies. Hippocrates (ca. 400 BC) recommended liver and honey as a cure for night blindness. Scurvy, the deficiency disease associated with lack of vitamin C (ascorbic acid), had been known since the long sea voyages of Vasco da Gama. Cartier in 1535 observed that Indians around Quebec cured the disease with extracts from the Annedda tree (*Thuya occidentalis*). The beneficial effects of citrus fruits were noticed as early as 1593 by Sir John Hawkins—"That which I have seen most fruitful for the sickness [scurvy] is sour oranges and lemmons." By the early years of the seventeenth century the East India Company was using lemon or lime juice in its merchant ships. The observations were formalized by Lind, a naval physician, in his *Treatise on Scurvy* (1753). He established that citrus fruits were very effective in preventing the onset of scurvy in people on long sea journeys and in curing their sore gums, loosened teeth, and little hemorrhages, the signs of the disease. The importance of lemon juice was recognized by the British admiralty ca. 1800 when it ordered its provision in ships to safeguard the health of the Channel fleet which had become endangered by epidemics of scurvy. When the fleet was based in the West Indies limes were used (not a very reliable source of vitamin C)—thus the term, "Limeys."

The direct link between disease and diet was first systematically examined by Eijkman in Java 1890-1897. Beri-beri was unknown in the Far East until the end of the nineteenth century when milling of cereals became popular. After the introduction of milling, the military hospital where Eijkman worked provided the patients with polished rice. They soon started to suffer from beri-beri (muscular weakness, peripheral neuritis, cardiovascular disturbances and often, later, massive edema). If the rice was not milled, so that the pericarp was retained, beri-beri was not evident. In the first controlled experiment on the induction of a vitamin deficiency in experimental animals Eijkman showed that if domestic fowls were fed polished rice they developed polyneuritis, a condition analogous to human beri-beri. If the birds were given an aqueous alcoholic extract of rice polishings, polyneuritis was no longer observed. Takaki prevented Japanese sailors from getting beri-beri by including sprouting barley in their provisions.

Eijkman's work was extended by Funk (1911) who found that the ingredient in rice polishings which cured polyneuritis in birds was an organic base. From this he proposed (1912) that all deficiency diseases except pellagra could be cured by the addition to the diet of "vital amines"—vitamin(e)s—which he thought were organic bases precipitable by phosphotungstic acid. Later work invalidated this generalization but retained the term "vitamin."

By 1906 it was appreciated, especially by Gowland Hopkins, that for animals to grow on a defined diet "accessory food factors" had to be provided in amounts that did not give significant increments in energy or protein intakes. Hopkins was by that time lecturer in chemical physiology in Cambridge. His observations on the importance of small amounts of milk in the diet, eventually published in the *Journal of Physiology* (1912). "Feeding experiments investigating the importance of accessory factors in normal dietaries" was enormously influential, leading to the award of a Nobel prize, jointly with Eijkman, in 1929. In his Chandler Medal address in 1922 Hopkins recalled "By this time I had come to the conclusion there must be something in normal foods which was not represented in a synthetic diet... the nature of which was unknown. Yet at first it seemed so unlikely. So much careful scientific work upon nutrition had been carried on for half a century or more—how could fundamentals have been missed? ... but ... the known fundamental foodstuffs ... had never been administered pure! ... moreover the unknown, although clearly of great importance, must be present in very small amounts ..." It is ironic that neither Hopkins nor any one else was able to repeat his classic experiments, probably because the amount of milk then given was less than that now thought necessary to provide an adequate vitamin intake for his animals. Moreover, although serving as the original source from which they would be isolated, milk is relatively low in B vitamins.

From 1912 work started to identify the accessory food factors in milk which were essential to life. Osborne and Mendel and McCollum and Davis showed there to be two classes of compound, called by McCollum the fat-soluble factor A and the water-soluble B component. Both A and B were soluble in alcohol. Purification of individual vitamins was a tedious process requiring a plentiful source of starting material which was relatively rich in the component of interest. Standard methods of isolation such as differential solvent extraction, precipitation and adsorption were the procedures generally used. The purification was followed by testing the fractions individually and in combination, until it had been established that only a single component was involved. Biological assays were used on animals, microorganisms, or occasionally (vitamin B_{12}) humans. Results were then compared with those of the starting material.

B Vitamins

The water-soluble component from milk was originally recognized because it contained the anti-beri-beri factor. It soon became clear that other nutritionally essential compounds were present, from which the anti-beri-beri factor could be distinguished by its instability above pH 8 or if heated at a pH above 5.5. Thiamine (vitamin B_1) was isolated by Jansen and Donath in 1926, its structure confirmed by synthesis by Williams and Cline in the Merck laboratories in 1936, and its biochemical function established by Peters and his colleagues (Chapter 5).

A peculiarity of thiamine is that the vitamin can easily become inactivated. An early instance was seen in 1941 when commercially reared mink became paralyzed (Chastek paralysis), a disorder which could be cured by giving the animals thiamine. The problem was traced to their having been fed fish that had partially decomposed. Later work showed that in decayed fish a microbial enzyme had been released, thiaminase, which destroyed the thiamine normally present in the food. A rather different process occurs when horses or cows are allowed to graze on bracken. This contains a protein which binds to thiamine, so reducing its availability. Once again the condition can be treated by administering the vitamin.

Pellagra was another widespread human deficiency disease whose signs were dermatitis and the breakdown of mucosa in the mouth and tongue. The vitamin involved was also claimed to be the anti-grey hair factor. In dogs its deficiency led to the condition known as "black tongue." Pellagra was most prevalent in African countries where people subsisted on diets rich in maize which is deficient in tryptophan, a precursor of nicotinamide. In 1915 Goldberger in the U.S. found that 6/11 convicts on a restricted diet developed pellagra, which was cured by giving liver, red meat or yeast extract. Nicotinamide was isolated from liver by Elvejhem in 1937 and found to be the anti-pellagra factor. It had by then been identified in coenzymes I and II (NAD^+ and $NADP^+$).

Riboflavin was another thermostable component in "vitamin B" required for rat growth. Its absence caused a form of dermatitis which was originally called rat pellagra because of its apparent similarity to human pellagra. When the molecule was isolated from milk it was found not to be nicotinamide but a new vitamin, a flavin, first called lactoflavin. The molecule was demonstrated by Kuhn, Gyorgy and Wagner-Jauregg to be closely similar to the prosthetic group in Warburg's "Old Yellow Enzyme" (see Chapters 4 and 5). Its structure was confirmed by synthesis in 1935 (Kuhn and Karrer). Because it contained an isoalloxazine ring linked to D-ribitol, it was renamed riboflavin and the dermatitis its deficiency produced in rats is now called acrodynia.

The last of the B vitamins to be identified in the water-soluble vitamin complex from milk was pyridoxine, vitamin B_6 (Birch and Gyorgy, 1936). This was needed to prevent a type of dermatitis in rats which was different from pellagra or acrodynia and could be accompanied by convulsions. Much of the early work on the mode of action of this vitamin came from experiments on microbial metabolism (Chapter 6).

Biotin, also a B vitamin, was isolated from duck egg yolks, crystallized by Kogl and Tonnes (1936), and found to be an essential growth factor for yeast. Its function as a vitamin in animals became evident from the work of du Vigneaud (1940). He was examining the effectiveness of ovalbumin as a protein source. When raw egg-white was used, puppies or rats failed to grow. The toxic properties of the egg-white were destroyed by heat, proteolytic enzymes, or HCl, suggesting they were due to a protein. The adverse effects could be overcome by biotin, and were caused by the presence in the raw egg-white of a glycoprotein, avidin, which specifically and very effectively bound biotin. Inhibition by avidin is now considered to be diagnostic for biotin-dependent reactions.

The final group of diseases accompanying B vitamin deficiencies were megaloblastic anemias arising from shortages of vitamin B_{12} (now cobalamin) or folic acid. Pernicious anemia was described by Addison in 1849. There is defective erythrocyte maturation so that large cells (megaloblasts) appear in the circulation. The patients often have atrophied gastric mucosa and may be achlorhydric (no HCl in the stomach). They may also show a staggering, ataxic gait, due to demyelination affecting the posterior columns and pyramidal tracts in the spinal cord.

In 1925 Whipple, examining the effects of repeatedly bleeding dogs, observed that the animals, not surprisingly, became anemic. The anemia responded very favorably when the dogs were given raw beef liver, so Whipple suggested that this might be helpful in the treatment of pernicious anemia. The treatment was independently introduced by Minot and Murphy in 1926. Patients were required to eat considerable amounts of raw liver. The megaloblasts in their circulation then declined, the reticulocyte count went up, followed by an increase in the number of erythrocytes and in the hemoglobin level.

Castle then showed (1929) that beef muscle was as effective as liver in preventing pernicious anemia, provided it was administered with normal gastric juice. He therefore concluded two factors were involved—an extrinsic one which was a component in liver or muscle and an intrinsic factor which was secreted by the stomach. Major efforts were therefore directed at identifying the extrinsic factor in liver or other meats.

One of the main difficulties in this work was that until the late 1940s the only means of assay for the active principle was to test the effect of the preparations on human patients. The disease could not be produced in animals. Patients with pernicious anemia are very sick. They also sometimes show spontaneous remissions so that monitoring the purification was extremely difficult. Nevertheless by 1945 patients were satisfactorily treated by taking less than 1 mg/day of active material compared with 400 g raw liver in 1926.

The solution to the assay problem came from the fortunate finding by Mary Schorb, then working in the poultry industry, of a microorganism, *Lactobacillus lactis dorner*, which required vitamin B_{12} for growth. With much quicker and more reliable assays the vitamin was isolated in 1948 in both the Merck and Glaxo laboratories. Its structure was determined by X-ray crystallography by Lenhert and Hodgkin (1961).

Microbial growth studies also gave an important clue to the intracellular role of vitamin B_{12} when it was observed that the presence of thymidine overcame the need for B_{12} in the culture medium of *Lactobacillus lactis dorner*, suggesting B_{12} was required for the biosynthesis of thymidine.

Intrinsic factor is a glycopeptide secreted by cells in the pyloric region of the stomach, which is needed for the translocation of the very large vitamin B_{12} molecule across the intestinal mucosal cell membranes.

The other megaloblastic anemia was described by Lucy Wills in patients in Bombay in 1931. The disease was induced in experimental animals by a diet of polished rice, white bread and the then known vitamins. Liver extracts were ineffective in restoring health but the sickness was reversed by giving autolyzed yeast preparations. In 1941, Mitchell, Snell and R.J. Williams isolated a growth factor for *Lactobacillus casei* from spinach foliage (folic acid). When tested in animals it prevented anemia and gave temporary remission of the megaloblastic anemia in patients with pernicious anemia. Folic acid is now known to be involved in the metabolism of 1C groups. It does not however affect the nervous disorder accompanying pernicious anemia, which can be halted by vitamin B_{12}.

Vitamin C

The discovery of plant and citrus extracts which cured scurvy has already been described. An inability to synthesize ascorbic acid is found only in primates and guinea pigs. That guinea pigs required an anti-scorbutic factor in their diet was shown by Holst and Frolich (1907). By 1924 a highly purified anti-scorbutic factor had been obtained from lemons. In unrelated studies not concerned with nutrition but with oxido-reducing systems a strongly reducing "hexuronic acid" was isolated by Albert Szent-Gyorgi, then in Gowland Hopkins' laboratory, from adrenal glands and citrus fruits. From its properties Szent-Gyorgi suggested the compound might be involved in biological oxidations but he did not know it was the anti-scorbutic factor, vitamin C. Szent-Gyorgi next visited the Mayo Clinic in the U.S. where, because of Kendall's work on the isolation of adrenocortical steroids, large supplies of adrenal glands were available from which hexuronic acid could be prepared. On his return to Hungary (1932-1933) Szent-Gyorgi carried with him 25 g of hexuronic acid. When a visitor, J. Swirbely, who was experienced in assaying vitamin C, joined his department in Szeged, Szent-Gyorgi gave him "hexuronic acid" to test; it was vitamin C (Szent-Gyorgi, 1963). Its structure was established by Haworth and the molecule renamed ascorbic acid. Crystalline vitamin C was prepared by Waugh and King in 1932 and its structure confirmed by synthesis the following year (Ault et al., 1933; Reichstein et al., 1933).

Fat-Soluble Vitamins

The identification of fat-soluble vitamins initially concentrated on the factor needed to prevent xerophthalmia in experimental animals. Xerophthalmia (dry eye) is a consequence of keratinization of the conjunctival and corneal surfaces of the eyes which may be followed by infection and ulceration. During World War I there was a high incidence of the complaint in children in Denmark where large quantities of butter were exported, and margarine used for home consumption. Restricting butter exports rapidly reduced the number of cases. It was soon observed (Osborne and Mendel) that cod-liver oil, a popular folk remedy, was a good source of "vitamin A." Some vegetables too were highly effective. Studies by Rosenheim and Drummond (1920) and Steenbock correlated vitamin activity with carotene content. This was questioned when it was shown by Karrer, Moore, and others that pale colored animal oils, such as halibut-liver oil, were very potent sources of vitamin A, but pure orange/yellow carotene was much less effective in supporting rat growth.

Night blindness has long been known as well as its rapid cure by ingestion of liver oils. In China in the seventh century AD night blindness was treated with pig or sheep liver. The rod cells of the retina are adapted to sensing light of low intensity. It is this function which is also impaired by vitamin A deficiency. The cells contain disc-like vesicles whose membranes have a high content of the light-absorbing protein, rho-dopsin. By the early 1940s George Wald at Harvard, who was to become another Nobel laureate in the vitamin field, had resolved the role of vitamin A in the visual cycle. Vitamin A aldehyde, retinal, was shown to be the prosthetic group in rhodopsin. From his work Wald could conclude: "Within the entire range of living organisms light sensitive structures contain carotenoids." The path from light reception to vision was thus opened up by Wald and is still an active field of biophysics.

The functions of vitamin A in the maintenance of epithelial cell integrity (Wolbach and Howe, 1925) emerged more slowly. An experiment by Fell and Mellanby (1953) with embryonic chick ectoderm cultures showed that in the presence of very high, non-physiological concentrations of vitamin A, the cells became mucous-secreting. In its absence keratinocytes developed. The active form of vitamin A in this case was probably retinoic acid, vitamin A acid, which it is now thought may promote differentiation by its effect on transcription from the genome.

Vitamin D

The nutritional experiments with carotene and fish oils led to the conclusion that a second fat-soluble compound was essential for normal rat growth. Rickets, the condition caused by vitamin D deficiency, is a disease afflicting children where, because of impaired calcification, bone formation is disturbed and the bones become bowed and otherwise deformed. In adults, especially multiparous women, vitamin D deficiency produced osteomalacia—demineralization of bone, leading to tenderness over the bones, pain, and muscle weakness. Rickets was particularly prevalent in slum areas. Glasgow, Vienna, and Lahore were notorious for the high incidence of the disease.

The key experiments leading to the identification of vitamin D were those of Mellanby (1918-1919) using puppies. When they were fed on bread, skimmed milk, linseed oil, yeast (to give B vitamins), and orange juice (vitamin C) the puppies developed rickets. When cod-liver oil and/or butter were added, rickets was prevented. The distinction between the effects of vitamin A and the anti-rachitic factor was aided by the sensitivity of vitamin A to oxidation. Aerated (oxidized) cod-liver oil no longer cured xerophthalmia but its anti-rachitic properties were unaffected (McCollum, 1922).

In the Artic Eskimos depended historically on fish for their supply of vitamin D, whereas in the tropics a supply is unnecessary. Excessive intakes of vitamins A and D can be lethal. The liver is the storage organ for fat-soluble vitamins; Eskimos avoided hypervitaminoses by discarding livers of polar bears which get a surfeit of vitamins A and D from their diet of seals and fish.

The importance of sunlight was demonstrated by Raczynsky (1912); a puppy reared normally in the light did not get rickets, whereas its litter-mate on the same diet, but brought up in the dark, did. The influence of ultraviolet (UV) light was dramatically shown by Chick and Dalyell in their classical work in Vienna after World War I. Large numbers of children were suffering from rickets. Milk by itself proved insufficient as a cure but if supplemented by cod-liver oil, or if the children were just taken into the sunlight, a cure was effected.

The effects of sunlight were explained when it was found that linseed or cotton-seed oils which were not good sources of the anti-rachitic factor, became much more effective after exposure to UV light. Similar experiments (Rosenheim and Webster) in which cholesterol or the plant sterol, ergosterol, were irradiated, showed that the products were anti-rachitic. By the early 1930s the structure of the active derivatives of cholesterol and ergosterol had been established by Windaus and his colleagues. Only since the 1960s has it become clear that the active form of vitamin D_3 is the 1,25-dihydroxy compound. This is believed to regulate levels of a protein affecting calcium uptake by intestinal mucosal cells and osteoblasts. Like steroids, the vitamin is thought to affect gene transcription.

Another fat-soluble vitamin, E, was found by Evans and Bishop in 1923. Pregnant rats on a defined diet (alcohol-extracted casein, cornstarch, and lard) supplemented with butter (vitamins A and D) and yeast extract (vitamin B group) produced few young because of fetal resorption. Male rats on the same diet were sterile. The disorders, which have not been identified in man, were corrected by wheat-germ oil, from which tocopherol, the active ingredient, was isolated in 1936. In spite of intensive investigations and a recognition that the vitamin is an antioxidant and destroyer of free radicals, the function of vitamin E remains obscure.

The last of the fat-soluble vitamins to be identified was vitamin K, found by Dam to be an anti-hemorrhagic factor for young chicks, distinct from vitamin C. Its structure was determined by Dam in collaboration with Karrer. Interest in the vitamin was intensified when it was discovered (Link, 1941) that dicoumarol, present in spoiled sweet clover, was the agent producing hypothrombinemia (giving prolonged blood-clotting time) in cattle. Since vitamin K is structurally similar to dicoumarol, the vitamin was presumptively implicated in thrombin formation. This has been fully substantiated by recent work on the role of vitamin K in the synthesis of prothrombin in the liver.

OTHER DIETETICALLY IMPORTANT FACTORS

The need to include a variety of minerals in experimental diets has already been mentioned; this was especially stressed (1920-1930) by Boyd-Orr, the director of the Rowett Institute for Animal Nutrition in Scotland. Increasingly refined food sources led to the identification of large numbers of trace elements (e.g., Cu, Mn, Mo, Zn) whose importance in the diet was suggested from hydroponic experiments with plant seedlings. Cobalt is an example of such a trace element. Vitamin B_{12} is synthesized by bacteria in the rumens of sheep and cattle but is absent from their fodder. In Australia, sheep feeding on cobalt-deficient pastures failed to thrive because vitamin B_{12} could no longer be made.

The last essential dietary components to which we will refer and which were also discovered through feeding experiments with rats, are certain unsaturated fatty acids identified as linoleic, linolenic, and arachidonic acids by Burr and Burr in 1930. The acids are required for the formation of complex lipids which are essential in membranes for the maintenance of their fluidity (Chapter 9). Deficiencies lead to a dermatitis which does not respond to additional B vitamin supplements or to oleic acid.

The greatly increased knowledge of nutrition in 1939 compared to that available in 1914 enabled a much better standard of health to be maintained for the whole population of the U.K. between 1939 and 1945 than in World War I (*How Britain was Fed in Wartime*, HMSO,1946). In 1936 Boyd-Orr had published "Food, Health and Income," drawing attention to the dietary inadequacies and poorer health of families in the U.K. on low incomes. Recommendations made there were very influential in determining amounts and the balance of food supplies between 1939 and 1945. Effecting the recommendations was greatly assisted by the appointment in 1940 of Jack Drummond, himself a nutritionist (see above), as scientific adviser to the Ministry of Food. This resulted in significant improvement in the nutritional value of the standard "white" loaf, calcium was added and the extraction rate of the wheat increased to 85% to improve the vitamin content of the flour. "British restaurants" were set up in many cities; these provided low-cost, nutritional meals without sacrificing food coupons.

GROWTH STUDIES WITH MICROORGANISMS

Microorganisms ("animalcules") had first been seen by van Leeuwenhoek (1676) in pepper-water which had been allowed to stand exposed to air. Similar experiments in the eighteenth century were not always successful because of difficulties the workers experienced in making simple lenses as effective as those used by van Leeuwenhoek. Joblot (1711) however noted that infusions of hay soon teemed with animalcules but that if the water was first boiled for 15 minutes and then covered it remained free from microorganisms for several days. This was soon confirmed by Spallanzani and made use of by Pasteur in his demonstration that lactic acid production in wine was due to contamination from microorganisms which could be collected by sucking air through a filter. From this it was a fairly straightforward step to the introduction, by the end of the nineteenth century, of sterilization by autoclaving. By then de Bary and Brefeld had shown how yeast cultures could be grown from single cells (cloned) a procedure adapted by Lister to give pure cultures of *Bacterium lactis.* An enormous stimulus to bacteriology came from experiments by Koch (1876) with anthrax bacilli which established that bacteria could cause disease. The birth of bacteriology necessitated the development of procedures for the reproducible culture of microorganisms. Among the requirements were suitable growth media.

The earliest microorganisms studied were pathogenic and had specialized and complex growth needs. Koch introduced the use of nutrient broths and agar slopes, which contained, for example, 0.5% of an enzymic digest of meat. Attempts were soon made to obtain completely defined media. For many microorganisms the presence of glucose, NH_4Cl, $MgSO_4$, K_2HPO_4, $FeSO_4$, $CaCl_2$, and trace elements proved sufficient for growth. In other cases adjuvants were needed. In 1934 R.J. Williams and Roehm showed that yeast growth was stimulated by the addition to the culture medium of vitamin B_1. Extracts prepared by boiling yeast cultures and filtering off the denatured proteins were also good sources of growth factors. The need to add yeast extract (cf. the addition of milk to rats' diets) prompted a search to define the factors in the extract which were essential for growth.

Pantothenic acid and biotin were thus found to be growth factors for yeast. Like riboflavin these molecules are incorporated into larger molecules in order to exert their essential metabolic function. Unlike the other vitamins there has been no evidence of pathological signs in man which can be attributed to dietary deficiencies in biotin or pantothenic acid.

In the 1930s streptococcal and staphylococcal infections were still major causes of death. Prontosil had been patented by I.G. Farben in 1932 and was first used successfully in Germany in 1933 to cure a case of staphylococcal infection in a child. Domagk continued the work, for which he received a Nobel prize in 1939. The active principle of prontosil is sulfanilamide. In the U.K. an MRC group led by Fildes was studying the nutrient requirements of a hemolytic streptococcus. D.D. Woods (1940) found the streptococcal growth was blocked by sulfanilamide. The inhibition could be reversed by yeast extract, from which the effective component—*p*-aminobenzoic acid—was isolated by Blanchard in 1941. Woods (1941) deduced that sulfa drugs were bacteriostatic because they competitively blocked the utilization of *p*-aminobenzoic acid. *p*-Aminobenzoic acid is part of folic acid, a growth factor for *Lactobacillus casei*, whose structure was identified in 1945 (Mitchell, Snell, and Williams). Woods' work identifying the basis of action of sulfonamides provided a logical approach to the development of new drugs (see Work and Work,1948).

Similar experiments by Lwoff showed nicotinic acid was an essential growth factor for *Hemophilus* bacteria and also for *Staph. aureus*. By the 1940s an aphorism had been coined: "Growth factors for microorganisms are B vitamins for higher animals."

The introduction of techniques for mutagenesis by UV irradiation or by the use of chemicals considerably extended the applications of microbial studies to nutrition (Davis, 1954-1955). Auxotrophic mutants were produced with nutrient dependencies not shown in the untreated parental strains (Beadle and Tatum,1940). The fortuitous discovery of penicillin by Fleming and its successful use in the treatment of infections (Florey) promoted exhaustive research into its mode of action. Eventually it was established that penicillin prevented the proliferation of gram-positive bacteria by blocking the synthesis of their cell walls

(Park, 1965; Strominger, 1965). This property of penicillin was exploited to facilitate the isolation of mutants. After exposure to a mutagen, cells were plated onto a nutrient deficient, penicillin-containing medium. Mutated bacteria could not then grow on the deficient medium. Unchanged, wild-type cells which could potentially grow, died as they were unable to synthesize their cell walls. The auxotrophic mutants could be recultured in the presence of the necessary growth supplements and used to analyze their metabolic roles (Lederburg and Zinder, 1948; Davis, 1954-1955).

METABOLIC DISEASES

Studies on energy needs and the identification of essential dietary components were performed mainly through observing animals, their heat production, respiration, and growth when the diet was inadequate. The existence of vitamins had been postulated, at least in part, from observations on, and treatment of, patients with scurvy, beri-beri, pellagra, etc. Further insights came from the study of humans with metabolic diseases. Metabolism is affected by exogenous and endogenous factors. Exogenous factors in the diet have been reviewed. The main endogenous influences are genetic or endocrine; it will be convenient to consider these as distinct although it is increasingly evident that in endocrine disorders there are often genetic factors.

Endocrine Diseases

Human dissection was legalized in Europe by the Emperor Fredric II in 1240. By the early sixteenth century thyroid, pituitary and adrenal "glands"—a term applied to many soft organs—had been described. "Glands" were frequently ducted (e.g. the pancreas, salivary, and bile ducts) but by 1766 von Haller had recognized the existence of ductless glands (thymus, thyroid, spleen) which "poured their special substances into veins and thus into the general circulation." By 1849, Berthold had demonstrated by extirpation and transplantation that the testis in cockerels produces a blood-borne substance conditioning sexual characteristics. Hormonal secretion was studied intensively from the mid-nineteenth century, particularly in the work of Claude Bernard on the role of the liver in glucose homeostasis, of Addison on the function of the suprarenals (adrenals), and of Brown-Sequard who employed glandular extracts, notably from testes, to correct hormone deficiencies. The term "hormone" was suggested by Sir William Hardy, and was used by Starling in 1905 to describe secretin which is released from the duodenum in response to the presence therein of dilute HCl and causes the pancreas to release its digestive enzymes.

By definition all hormones affect the behavior of their target cells. Examples of the interplay between endocrine disturbances and their biochemical consequences are provided by some of the diseases of the thyroid, which directly affects basal metabolic rate, and diabetes mellitus, where glucose metabolism is deranged.

The Thyroid Diseases

Goiter, the enlargement of the thyroid glands in the neck, was recognized in antiquity. Ancient Egyptian tomb carvings show people with pendulous swellings in the neck. In the sixteenth century Leonardo da Vinci drew goitrous subjects and in *The Tempest* Shakespeare refers to "Mountaineers dew-lapped like bulls, whose throats had hanging at them, wallets of flesh."

The prevalence of goiter in isolated mountainous regions such as the Alps, the Pyrenees, and Derbyshire was also noted. The Chinese are credited with the earliest uses of dried sponges or seaweed ground up in wine to alleviate the condition. Marco Polo (1241) wrote that goiters in China were thought to be occasioned by the nature of the water people drank. The link between goiter, retarded development, and congenital idiocy was accepted by the time of Paracelsus (1493-1541). That thyroid insufficiency could afflict people, especially women, in middle age (myxedema), was not recognized until the 1870s when physicians at Guy's Hospital in London described patients who increased in weight, showed facial thickening and became languorous and placid. Exopthalmia, the protrusion of the eyeballs, was mentioned in Persian literature (1136 AD), and the classical description of exopthalmic goiter associated with hyperthyroidism, where protruding eyes are accompanied by rapid heart beat and excessive nervous activity, was given by Parry (1755-1822) and Graves (1796-1853).

In 1858 Schiff removed the thyroid gland from animals and found they could not survive. From his, and later, Horsley's, experiments and from the clinical observations it was concluded that the effects of removal or damage to the thyroid were due to a loss of its internal secretions. Confirmation of this came when patients with myxoedema were successfully treated by thyroid extracts or even by eating thyroid tissue.

Iodine was isolated from seaweed in the early nineteenth century by Courtois. It is thought that Prout in 1816 was one of the earliest to administer iodine to patients, having first tried it out on himself, in an attempt to cure thyroid deficiency. Toxic effects were soon reported following administration of elementary iodine, but once it had been demonstrated that endemic goiter occurred where amounts of iodine in drinking water and the soil were low, the use of iodized table salt was recommended to prevent goiter in mountainous regions of France and elsewhere. By the end of the century Baumann had shown that the active constituent in thyroid secretion was an iodinated organic compound. Its isolation from 3 tons of pig thyroid by Kendall (1914) required some procedure by which the purification could be followed. Biological assays were devised, based on the stimulation of the metabolic rate produced by thyroid extracts, first observed by Magnus-Levy in 1895, or on the promotion of metamorphosis in tadpoles which was reported a few years later by Gudernatsch. By 1926 the structure of thyroxine had been established by Harrington; it was synthesized by Harrington and Barger the following year. A further active component, triiodothyronine (T3) was identified by Gross and Pitt-Rivers in 1952.

Demonstration that the effects of thyrotoxicosis or goiter are due to an excess or deficiency in (thyroxine + T3) secretion does not explain how the diseases originate nor why development and metabolic rate are affected. It is thought that some cases of thyrotoxicosis (Graves' disease) may be caused by abnormal immune responses mimicking the effects of thyroid-stimulating hormone on the thyroid gland.

Thyroxine and T3 are believed to have regulatory effects on the genome; how these cause the metabolic rate to be increased still remains a mystery.

Diabetes Mellitus

Diabetes mellitus was the name given to the condition in which glucose was excreted into the urine, which was sweet to the taste and markedly increased in volume. The sweet taste of the urine was described by Indian physicians in the seventh century. Dobson in 1715 attributed this to the presence of sugar which was shown by Peligot (1838) to be glucose. A quantitative test for glucose was devised by Fehling in 1848 based on a reaction described by Barreswill (1817-1870), a friend of Bernard. Glucose and other "reducing sugars" with free putative aldehyde or ketonic groups reduced blue alkaline cupric tartrate to give a red-brown precipitate. Juvenile Onset Diabetes, Type I, is that form of the disease which becomes evident in childhood and is due to an insufficiency in the production of insulin.

Willis (1621-1675) appreciated that the sweetness of urine in the "pissing-evil" must be preceded by sweetness in the blood. Various surgeons in the seventeenth century explored the effects of extirpating the pancreas from dogs. Any dog which remained alive is unlikely to have had its pancreas completely removed, but reports on the survivors refer to polydipsia and polyuria, frequently observed in untreated diabetics.

Pancreatic islets were described by Langerhans; they were more evident in fetal than in adult tissue. Claude Bernard, the pioneer in the study of secretion and the function of the liver in the control of blood sugar, died in 1877. His studies were continued by younger French colleagues who linked diabetes both with a failure of an internal secretion and with lesions in the pancreas. By 1900 these ideas had crystallized to the view that the islets of Langerhans were responsible for the production of an internal secretion affecting blood glucose levels. Direct evidence for this had been provided by von Mering and Minkowski (1890-1893) who successfully extirpated the pancreas from dogs and induced diabetes. The livers contained negligible amounts of glycogen and if the dogs were given glucose they excreted it in the urine together with the "ketone bodies," acetoacetate and β-hydroxybutyrate. By this time these had been described in the urines from human diabetics. In extremely important experiments, exemplifying what became a standard approach to endocrinology, Hedin in 1893 and Minkowski in 1908 further showed that a pancreatic graft in a pancreatectomized dog prevented glucosuria. The effects of extirpating the gland had been corrected by replacing the deleted tissue. If the graft was removed, glucosuria recommenced.

The race was then on to isolate the secreted agent from the pancreatic islet cells. Juvenile diabetes was a disease from which death inexorably followed in teenage or the early 20s. Heroic attempts had for a long time been made to control the condition by diet; by the early part of this century low carbohydrate, high protein-fat diets were favored. Pancreatic extracts administered by mouth were inevitably ineffective since insulin is destroyed by digestive enzymes. A number of workers isolated insulin in various stages of purity including Zuelzer (1908) who obtained a preparation which dramatically lowered blood glucose; however, it was extremely impure, producing such serious side-effects that the study was not pursued. Gley and more effectively Paulesco, a Rumanian physiologist, also obtained insulin, but it wasn't until 1921 that reproducible production of highly purified insulin and its successful administration to patients was achieved by the Toronto group of Banting, Best, Collip, and Macleod.

The origins of diabetes mellitus are still being investigated. There is a familial trait—certain histocompatability phenotypes and perhaps other non-HLA genes are more frequently displayed by juvenile diabetics than others. Viral infections in childhood may precipitate immune responses which damage the β islet cells. Other types of diabetes, such as that shown by middle-aged or older patients, have different causes and can often be controlled by appropriate diet.

The promotion by insulin of glucose uptake by muscle and fat cells (adipocytes), of glycogen deposition in liver and muscle, and its stimulation of growth soon emerged as the purified hormone became available for study. Although insulin was crystallized by Abel in 1926, its primary structure established by Sanger in 1953 (see Chapter 10), and its shape determined crystallographically by Hodgkin in 1966, the molecular basis of its action is still uncertain and controversial. Clarification of pathways, especially the interactions between fat and carbohydrate oxidation (see Chapters 4 and 5), has however been greatly helped by examining metabolic changes in diabetics or in animals in which diabetes has been induced by drugs such as streptozotocin.

GENETIC DISEASES

Archibald Garrod can be regarded as the father of medical genetics. He qualified from St. Bartholomew's hospital, London, in 1884. By 1904 he had become Physician-in-Charge of the Children's Department, combining his clinical responsibilities with a pioneering interest in the molecular basis of disease, reflected in the lectures he gave on chemical pathology. Early in this century he coined the phrase, "Inborn Errors of Metabolism," to describe certain diseases which showed Mendelian patterns of inheritance and which were associated with reduced activities of enzymes. His later book, *The Inborn Factors in Disease* (1931), published only five years before his death, added gout, Gaucher's disease (a disorder in complex lipid metabolism), hemophilia, and porphyria to his original list of diseases due to inborn errors. In spite of the respect in which Garrod, by then Regius Professor of Medicine in Oxford, was held, his recognition of inherited, genetically based illnesses was unappreciated by contemporary physicians.

One reason for this was that some of the conditions had relatively trivial consequences for the patients. In other cases, like porphyria or hemophilia, it would be another 20 to 30 years before the biochemical abnormality could be identified. Furthermore clinical interests at the time largely focused on diseases with extrinsic origins—bacterial or viral infections—when constitutional idiosyncracies appeared of little relevance.

One of the classical studies made by Garrod was of alcaptonuria. Here the abnormality, in which the urine turns black soon after voiding, although disconcerting, has only slight effects on the patient. The compound responsible is homogentisic acid. Phenylalanine, tyrosine, and 4-hydroxyphenylpyruvate are normally metabolized through homogentisic acid which is oxidized in the liver by homogentisic acid oxidase and other enzymes, finally yielding fumarate and acetoacetate. In alcaptonurics, homogentisic acid oxidase is missing. Phenylalanine and tyrosine in the diet are therefore incompletely oxidized and homogentisic acid passes out into the urine where it spontaneously oxidizes to a black quinonoid compound.

A much more serious genetic disease, first described by Folling in 1934, is phenylketonuria. Here the disturbance in phenylalanine metabolism is due to an autosomal recessive deficiency in liver phenylalanine hydroxylase (Jervis, 1954) which normally converts significant amounts of phenylalanine to tyrosine. Phenylalanine can therefore only be metabolized to phenylpyruvate and other derivatives, a route which is inadequate to dispose of all the phenylalanine in the diet. The amino acid and phenylpyruvate therefore accummulate. The condition is characterized by serious mental retardation, for reasons which are unknown. By the early 1950s it was found that if the condition is diagnosed at birth and amounts of phenylalanine in the diet immediately and permamently reduced, mental retardation can be minimized. The defect is shown only in liver and is not detectable in amniotic fluid cells nor in fibroblasts. A very sensitive bacterial assay has therefore been developed for routine screening of phenylalanine levels in body fluids in newborn babies.

There are now thought to be several thousand different genetic diseases, about 10% of which have known biochemical lesions. As has already been seen with the thyroid diseases and diabetes, the phenotypic manifestation, hemophilia, for example, may have genetically, biochemically or clinically different causes. Some of the biochemically identified disturbances, such as those affecting glycogen or galactose, have been important in establishing metabolic pathways (see Chapter 4).

The experiments described in this chapter show how three fairly simple approaches—respiratory studies, balance measurements, and observations on growth—were applied to animals including man, so that by 1930-1940 energy requirements and the major and essential minor dietary constituents were known. In a few instances of dietary deficiencies, genetic, or endocrine disorders, urinary analyses indicated alterations in excreted products which shed light on normal metabolic routes. Often, however, the signs of the illness were not easily related to the abnormal catabolites. What was needed next was to determine how oxidation of foodstuffs was brought about, how and in what form the energy was released and utilized by the cells for chemical and physical work, and what were the intracellular functions of the essential food factors and trace elements. This information could only be obtained by investigating chemical reactions taking place in isolated cells and tissues, which could also be shown to occur in the intact animal.

REFERENCES

Albanese, A.A., Ed. (1959). Protein & Amino-Acid Nutrition. Academic Press, New York.

Bodansky, M. & Bodansky, O. (1952). Biochemistry of Disease, 2nd ed. Macmillan, New York.

Boyd-Orr, J. (1936). Food, Health and Income. Macmillan, London.

Cartier, J. (reprinted 1953) La Grosse Maladie. 19th Congres International de Physiologie, Montreal.

Davis, B.D. (1954-1955). Biochemical Explorations with Bacterial Mutants. Harvey Lectures L, 230-257.

Draznin, B., Melmed, S. & LeRoitt, D., Eds. (1989). Molecular and Cellular Biology of Diabetes Mellitus. Alan R. Liss, New York.

Fell, H.B. & Mellanby, E. (1953). Metaplasia produced in cultures of chick ectoderm in high vitamin A. J. Physiol. 119, 470-488.

Fieser, L.F. & Fieser, M. (1950) Organic Chemistry, 2nd ed. D.C. Heath & Co., Boston.

Galjaard, H. (1980). Genetic Metabolic Diseases. Elsevier North Holland, Amsterdam.

Garrod, A.E. (Reprinted 1963, Harris, H. Ed.). Inborn Errors of Metabolism. Oxford University Press.

Garrod, A.E. (Reprinted 1989, Scriver, C.R. & Childs, B. Ed.) The Inborn Factors in Disease. Oxford University Press.

HMSO (1946). How Britain was fed in Wartime.

Iason, A.H. (1946). The Thyroid Gland in Medical History. Froben Press, New York.

Jensen, H. (1948). The Internal secretion of the pancreas In: The Hormones (Pincus, G. & Thimann, K.V. Eds.), Vol. 1, pp. 301-302. Academic Press, New York.

Knight, B.C.J.G. (1945). Growth Factors in Microbiology Vit. & Hormones 3, 105-228.

Lusk, G. (1928). The Science of Nutrition. W.B. Saunders, Co. Philadelphia & London.

Medvei, V.C. (1982) A History of Endocrinology. MTP Press, Lancaster, U.K.

Needham, J. & Baldwin, E. Eds. (1949). Hopkins and Biochemistry, 1861-1947. Heffer & Sons, Ltd. Cambridge, U.K.
Passmore, R. & Eastwood, A. (1986). Human Nutrition and Dietetics. Churchill Livingstone, Edinburgh.
Rose, W.C. (1938). The nutritive significance of the amino-acids. Physiol. Rev. 18, 109-138.
Stanier, R.Y., Doudoroff, M., & Adelberg, E.A. (1958). General Microbiology. Macmillan, London.
Stephenson, M. (1949) Bacterial Metabolism. 3rd ed. Longmans, London.
SubbaRow, Y., Baird-Hastings, A., & Elken, M. (1945). Vitamin B_{12}. Vitamin & Hormones 3, 238-296.
Szent-Gyorgi, A. (1963). Lost in the twentieth century. Annu. Rev. Biochem. 32, 1-14.
Wald, G. (1943). The photoreceptor function of the carotenoids and vitamins A. Vit. & Hormones 1, 195-227.
Waxman, D.J. & Strominger, J. L. (1983). Penicillin-binding proteins and the mechanism of action of ß-lactam antibiotics. Annu. Rev. Biochem. 52, 825-869.
Wolbach, S.B. & Howe, P.R. (1925). Tissue changes following deprivation of fat-soluble A vitamin. J. Exp. Med. 42, 753-778.
Work, T.S. & Work, E. (1948). Basis of Chemotherapy. Oliver & Boyd, Edinburgh.

Chapter 4

CARBOHYDRATE UTILIZATION:
GLYCOLYSIS AND RELATED ACTIVITIES

INTRODUCTION

The term "glycolysis" was introduced by Lepine (1909) to describe the disappearance of carbohydrate during metabolism. It is now used for the breakdown of glycogen or glucose that occurs anaerobically through pyruvate to give lactic acid in liver and muscle or ethanol in yeast. The glycolytic route was the first biochemical pathway to be established; all its enzymes have been sequenced and many of their three-dimensional structures determined, as well as mechanisms whereby the enzymes catalyze the different reactions. Identification of the sugar phosphate esters involved in the pathways led directly to the discovery of phosphocreatine, ATP and the realization of the importance of ATP as the energy currency of cells (Lipmann,1941).

Two other features associated with glycolysis will be considered. The breakdown of liver glycogen and release of blood glucose observed by Claude Bernard formed the basis of his ideas on glucose homeostasis; by the 1950s detailed analysis of these reactions and of the formation of liver glycogen provided the groundwork for the explosive and continuing development of molecular endocrinology.

Microscopists in the nineteenth century had begun to describe changes in the appearance of muscle fibers during contraction. Their experiments were concurrent with those of physiologists examining the relation between the work done by striated muscle and its heat production. Studies of glycolysis in muscle and the investigation of muscle contraction provided an early example of the strengths and difficulties of interdisciplinary collaboration.

DEVELOPMENT OF ANALYTICAL TECHNIQUES

The discovery (see below) by the Buchners that cell-free preparations from yeast were able to ferment glucose, was exploited by Harden and Young who showed that inorganic phosphate was an essential component in the fermentation process. Analysis of the glycolytic intermediates required procedures by which different phosphorylated sugars could be identified and estimated (Umbreit et al.,1945). Enzymes had first to be inactivated and proteins removed, usually at 0-4 °C. Precipitation at 5-10% trichloroacetic acid was frequently employed. If the extract contained significant glycogen, it was removed by precipitation at 50% ethanol, so that it did not interfere with subsequent steps in the isolation. The sugar phosphates were then converted to their barium salts which were separated by differences in their solubilities at neutral or acid pH in the presence or absence of ethanol.

Sugar phosphates were digested and the inorganic phosphate produced, estimated gravimetrically as phosphomolybdate. After Fiske and SubbaRow had found that phosphomolybdate could be reduced to molybdenum blue (1925), the phosphate was estimated colorimetrically. The Duboscq colorimeter was the first to be generally available; here light transmissions by standard and unknown solutions were matched by eye. Gelatine filters were sometimes used to give approximately monochromatic light.

By the 1940s spectrophotometers had been designed with photoelectric cells replacing visual comparison; prisms were incorporated to give incident light of defined wavelength. Ultraviolet (UV) light sources which could provide a range of wavelengths from 230 to 400 nm replaced earlier mercury vapor lamps which had restricted emissions. Spectrophotometry could thus be extended into the UV for the determination of large numbers of metabolites, especially nucleotides and proteins which have no measurable absorption in the visible region. Reduced pyridine nucleotide coenzymes (NADH and NADPH) were distinguished from their oxidized forms (NAD^+ and $NADP^+$) by their absorption at 340 nm (Warburg and Christian, 1936). As more and more enzymes were identified and became available commercially, methods of estimation were increasingly based on the coupling of enzyme reactions whose endpoint was the production or utilization of NAD(P)H. A further increase in sensitivity was obtained if the fluorescence of the pyridine nucelotides was measured rather than their absorption (Chance, 1962).

In 1928 Lohmann found that some phosphate esters were unstable in M HCl with inorganic phosphate being released after 7-15 minutes at 100°C, thus becoming measurable by the Fiske and SubbaRow method. Later investigations showed that in the glycolytic pathway phosphate present in acid anhydride bonds, (e.g. β and γ groups of ATP, or the C1 positions of glucose), were the main contributants to acid-labile P. Phosphate on the 5-position of pentoses or the 6-position of glucose and fructose and in 2- or 3-phosphoglyceric acid was stable and required ashing for complete conversion to inorganic phosphate. These procedures enabled the different phosphate esters in glycolytic and related pathways to be identified. Separation of phosphate esters by paper or column chromatography was developed in the 1950s (see Chapter 10), to be followed a little later by thin layer chromatography. These greatly increased efficiency, sensitivity and speed of the analytical processes.

THE GLYCOLYTIC PATHWAYS

Yeast

The breakdown of glucose by yeast to give ethanol, acetic acid, and carbon dioxide was examined quantitatively by Lavoisier (1789) and Gay-Lussac (1810). From his studies (Chapter 2) Pasteur described fermentation as "life without air", attributing the process to the presence of yeast cells whose effects were dependent on the vital force. The first suggestion that an "unorganized" ferment was responsible for fermentation was due to Traube (1858). Support for his ideas came from Berthollet (1860) who extracted yeast with water, precipitated the extract with alcohol, and found that the redissolved precipitate could still cause inversion of sucrose. Berthollet concluded "The living being is not the ferment but that which engenders it."

Edouard Buchner became interested in fermentation while a Privat-Dozent in Munich. He made use of a method described in 1846 whereby yeast could be disrupted by grinding it in a mortar with fine sand. His studies were of little interest in the laboratory where he was working. However, in 1894 his elder brother, Hans, who was appointed to the Chair of Hygiene in Munich, needed to obtain protein from yeast for his work in immunology. Attempting to extract the proteins by grinding them with kieselguhr gave preparations which were difficult to preserve. Hans recalled that sugars were used to preserve fruit, and suggested to Edouard, who had joined him for a vacation, that 40% glucose and other preservatives should be added to the yeast extract. Edouard saw that fermentation continued in the extract even if antiseptics were added to the preparations. He concluded (1897): "First, it is proved that to bring about the fermentation process, such a complicated apparatus as represented by the yeast cell is not required. . . . the bearer of the fermenting action of the press juice is more truly a dissolved substance, doubtless a protein; this will be designated as a zymase." (See Teich, 1992).

These observations established that fermentation could occur in yeast extracts in the absence of the cells themselves. The vitalist view that cell activities required the presence of organized cells was therefore no longer tenable. Edouard ascribed the breakdown of glucose to the presence in the extracts of ferments—"zymase." Cell-free fermentation was reported in 1897 but was not immediately accepted either by believers in protoplasm (see Chapter 9) or by brewing technologists who were by now firmly in the Pasteur tradition, and initially reported their inability to repeat Buchner's findings. Scepticism did not last long and by 1898 Edouard had been made Director of the Institute for the Fermentation Industry. He was awarded a Nobel Prize in 1907. That glycolytic enzymes are soluble greatly simplified separation and identification of the substrates and enzymes concerned in the pathway.

Another cell-free yeast preparation which was quite widely employed was "lebedevsaft" (1911-1912). Here yeast was macerated with water at 25-35 °C for 2 days before being filtered. A thin layer of toluene was used to prevent contamination by airborne organisms. Activities in these extracts declined very fast, the decay initially being attributed to proteolysis. To prevent this fall-off in activity Harden, working with Young in the Lister Institute, added boiled, filtered, autolyzed yeast juice to the fresh fermenting yeast system. Fermentation was dramatically enhanced, though not by inhibition of proteases but because of the presence in the autolyzed juice of a heat-stable, alcohol-soluble factor they called co-zymase (1904-1905). Because of its thermal stability they concluded co-zymase was unlikely to be a protein.

If fresh yeast extracts were ultrafiltered two fractions were obtained, a yellow solution, and a residue containing protein and glycogen. Neither of these broke down glucose by itself but when recombined, alcohol production was restored. Inorganic phosphate (P_i) was recognized by Harden and Young as an essential component in the ultrafiltrate. The need for inorganic phosphate had been noted by Wroblewski (1901) after he had observed the stimulation of fermentation by P_i. The phosphate was incorporated into organophosphate (Ivanow, 1905). In the next 20 years hexose diphosphate [fructose 1,6 bisphosphate, (F-1,6 bisP)] and hexose monophosphates [a mixture of glucose and fructose 6-phosphates, (G-6-P and F-6-P)] (Harden and Robison) were successively isolated via their barium salts. When reintroduced into cell-free yeast systems, the sugar phosphates were broken down to give ethanol.

In 1908 Harden and Young proposed that fermentation and phosphorylation were separate, coupled, reactions whereby the esterification of one sugar molecule by inorganic phosphate induced the decomposition of another molecule of glucose to carbon dioxide and alcohol. Harden maintained this view throughout his life (1929, Nobel lecture), never accepting that the phosphorylated compounds were intermediates in fermentation. The importance of phosphorylation was questioned by others, especially Neuberg, who thought it occurred only in damaged cells (dried, ground, or otherwise macerated yeast preparations) which, unlike intact yeast, could ferment added phosphate esters. Phosphorylated compounds cannot enter fresh yeast cells.

In contrast to experiments which attempted to identify phosphorylated intermediates on the "look and see what is there" basis, others were designed to investigate possible pathways of glucose breakdown suggested from its known chemical reactivity. Treating glucose *in vitro* with alkali caused the formation, *inter alia*, of methyl glyoxal (pyruvaldehyde), which could be isolated as its osazone. This product could also be derived from glyceraldehyde, dihydroxyacetone, and lactic aldehyde. Treating methyl glyoxal with alkali gave rise to lactic acid. Buchner and Meisenheimer therefore proposed (1909) that methyl glyoxal, glyceraldehyde, or dihydroxyacetone might be precursors of lactic acid.

Experiments with glyceraldehyde and dihydroxyacetone showed them to be fermentable, but results with methyl glyoxal were conflicting. With Lebedev juice no lactic acid was formed, but with top yeast Neuberg reported that lactic acid was detectable. In 1913 he and Kerb therefore proposed sugar was converted to methyl glyoxal from which pyruvic acid and ethanol were derived:

$$CH_3COCHO \rightarrow \text{pyruvic acid} \rightarrow CH_3CHO + CO_2$$

methyl glyoxal carboxylase

$$CH_3COCHO + CH_3CHO \rightarrow CH_3COCO_2H + C_2H_5OH$$

Cannizzaro reaction

Glyoxalase, the glutathione-dependent enzyme which catalyzed the conversion of methyl glyoxal to lactic acid, was isolated by Neuberg and by Dakin and Dudley.

Muscle

Complementary experiments with muscle also showed that glycogen breakdown required inorganic phosphate. The formation of lactic acid from glycogen in muscle was of particular concern to physiologists. Metabolic changes in this tissue proceed very rapidly and can be induced simply by handling. Fletcher and Hopkins (1907) found that if small muscles were selected they could be cooled very rapidly by dropping them into ice-cold alcohol. Under these circumstances glycogen was still detectable in fresh muscle but had diminished if the muscle had been exercised, with a concomitant rise in lactic acid. Amounts of lactic acid were proportional to the amount of work done.

Physiologists therefore suggested that lactic acid or the increase in H^+ concentration caused muscle contraction.

The Fiske and SubbaRow method for determining P_i has already been mentioned. Color development required several minutes to achieve maximum intensity. While examining P_i concentrations in muscle extracts, Grace and Philip Eggleton observed (1927) that the color continued to increase after the time normally allowed for the reduction of phosphomolybdate. They concluded that an acid-labile phosphate compound was present—a phosphagen. They and Fiske and SubbaRow independently isolated the molecule and found it to be phosphocreatine (PC). The following year Lohmann, working in Meyerhof's laboratory, showed the presence of a second acid-labile compound in muscle extracts which was rather more slowly broken down in acid-molybdate and was a pyrophosphate. Muscles exercised to exhaustion or in rigor contained virtually no pyrophosphate but had a marked increase in P_i. Lohmann then showed that AMP in muscle was a breakdown product of the triphosphate, ATP.This discovery of ATP was described at the 13th Congress of Physiology in Boston in 1929. Although there were a number of papers devoted to muscle physiology, at least some of the accounts of the meeting make no mention of this seminal finding. Fiske and SubbaRow reported the isolation and crystallization of ATP from muscle in the same year.

Identification of the energy source for muscle contraction and determination of the order in which the phosphate esters were metabolized was helped by the use of inhibitors. These inhibitors blocked different stages in glycolysis and caused preceding substrates to accumulate in quantities which could greatly exceed those normally present. The compounds were then isolated, identified, and used as specific substrates to identify the enzymes involved in their metabolism. Iodoacetic acid (IAc) was one of the most important inhibitors used to analyze glycolysis.

Lundsgaard (1930) was originally concerned with a study of the specific dynamic action of amino acids, i.e. the stimulation which certain amino acids in the diet gave to energy output. He used various iodinated amino acids including, as an analogue for iodoglycine, iodoacetate. This he injected into frogs and rabbits. Muscle contraction continued for a while but then stopped, rigor having developed. The muscles were arrested in a highly contracted state. When this was repeated with frog gastrocnemius muscle *in vitro,* Lundsgaard observed that no lactic acid was produced, indeed the pH of the preparation actually rose, i.e. neither lactic acid nor H^+ were available to cause the contractions. Phosphocreatine however had disappeared, and P_i levels increased. Hexose mono- and diphosphates also increased. The theory that muscle contraction was initiated by lactic acid or by the increase in H^+ concentration therefore had to be abandoned, with phosphocreatine taking over the central role (see below) to be displaced by ATP after the discovery of the Lohmann reaction:

$$\text{Phosphocreatine} + \text{ADP} \rightleftharpoons \text{ATP} + \text{Creatine}$$

In 1931 Lundsgaard suggested that the energy used in muscle contraction was derived from phosphate bond energy supplied by glycolysis or respiratory oxidation.

The basis of the action of iodoacetate on muscle contraction was uncovered by Dickens and Rapkine (ca. 1933). They found iodoacetate alkylated SH groups on proteins, especially those in glyceraldehyde 3-phosphate dehydrogenase (G-3-PDH). When the enzyme was inhibited precursors accumulated—hexose mono- and diphosphates—as in Lundsgaard's experiments.

The natural substrate for the dehydrogenase, glyceraldehyde-3-phosphate (G-3-P), had been synthesized earlier by Hermann Fischer, Emil Fischer's son, and Baer in 1932. In 1934 Meyerhof and Lohmann synthesized hexose diphosphate, establishing it to be fructose 1,6 bisphosphate (F-1, 6 bis P). With F-1,6 bisP as substrate and hydrazine to trap the aldehydic and ketonic products of the reaction, G-3-P was identified in the mixture of G-3-P and dihydroxyacetone phosphate which resulted. Triose phosphate isomerase was then isolated and the importance of phosphorylated 3C derivatives established.

Mechanism of Action of Glyceraldehyde 3-Phosphate Dehydrogenase

Although Harden and Young had observed in 1904 that a heat-stable factor, cozymase, was essential for glycolysis in their yeast preparations, it was some time before similar conclusions were reached by muscle physiologists. Neuberg (1913) found that aged muscle extracts reactivated yeast preparations and Embden and colleagues (1914-1915) showed that fresh yeast extracts contained something which greatly stimulated the utilization of glucose by muscle extracts. This fraction was resolved into two "coenzymes": one was identified by Meyerhof and Lohmann as a phosphate-transferring coenzyme: (i.e., ATP), and the other by von Euler (1923) as an adenine nucleotide-containing compound. Some workers, especially Warburg, disputed von Euler's identification and suggested the coenzyme was contaminated with AMP or ATP. Warburg had shown that the hydrogen-transferring coenzyme from red blood cells was a dinucleotide of adenine and nictotinamide containing three molecules of phosphate, which he called triphosphopyridine nucleotide, (TPN). This was later renamed coenzyme II and then $NADP^+$. In 1936 von Euler and Warburg both concluded that cozymase had adenine, nicotinamide, and two phosphorus atoms per molecule, thus DPN, coenzyme I or NAD^+. A change in the spectrum on reduction was shown to be due to the quaternary N of the pyridinium structure changing to the tertiary structure of dihydropyridine.

G-3-PDH was crystallized in 1939 by Warburg and Christian. The enzyme from yeast bound NAD^+ very strongly; that from horse liver contained NAD^+ which could be removed if the enzyme was passed over charcoal. A second enzyme was found to be present in the dehydrogenase system, 1,3 diphosphoglycerate kinase, which transferred the very unstable phosphate group from the mixed anhydride, 1,3 diphosphoglyceric acid, to ADP to yield ATP. When purified G-3-PDH was used, the reaction required P_i and NAD^+. The overall pathway for the reaction was thus established, involving an oxidative step, the uptake of P_i and the formation of a molecule of ATP. If arsenate was added, it substituted for phosphate and the mixed acid anhydride spontaneously broke down, preventing the substrate level phosphorylation.

Once high concentrations of pure crystalline rabbit muscle G-3-PDH became available, it became possible to demonstrate photometrically that the addition of G-3-P reduced NAD^+ bound to the enzyme. *p*-Chlormercuribenzoate, another compound which reacts with SH groups, displaced bound NAD^+ from the enzyme and inhibited its action. Free glyceraldehyde or acetaldehyde was then used as substrate as they were utilized by the enzyme but much less rapidly than G-3-P. With high (stoichiometric) amounts of pure enzyme, Racker and Krimsky (1952) found acetaldehyde was oxidatively phosphorylated by G-3-PDH to acetyl phosphate. By analogy with reactions on acetyl coenzyme A (see chapter 5), Lynen and Reichert suggested the essential thiol (SH) group in G-3-PDH formed a thioester with an acyl intermediate. SH groups could be titrated amperometrically with iodosobenzoate. In the presence of the substrate, G-3-P, 2 out of 5 SH groups on the protein did not react with the reagent and so were considered to be in the catalytic site of the enzyme. Interaction of alkylating agents with SH was confirmed spectrophotometrically and a mechanism for the reaction postulated.

By the late 1920s it was generally recognized that the breakdown of hexose phosphate in glycolysis yielded pyruvate. Enzymes catalyzing the steps from 1,3-diphosphoglyceric acid were identified principally from Embden's, Lohmann's and Meyerhof's laboratories. One further inhibitor, fluoride, was used, which had been found by Embden to inhibit phosphatases. In the presence of fluoride and F1,6 bisP, muscle minces accumulated phosphoglyceric acids, 2- and 3-PGA. Embden and his colleagues next showed (1933) that muscle extracts could convert phosphoglyceric acid to pyruvate and P_i. Phosphopyruvate was identified by Lohmann and Meyerhof as an intermediate. Fluoride very effectively inhibited the conversion by enolase of 2-PGA to phosphoenolpyruvate (Meyerhof et al., 1935-1938).

Methyl glyoxal was finally removed from the glycolytic pathway after Lohmann showed (1932) that dialysed muscle extracts, supplemented with P_i, ATP, NAD^+, etc., converted glycogen to lactic acid. If glutathione was not present to reactivate glyoxalase, methyl glyoxal did not form lactic acid.

It is difficult to realize that there were still unknown reactions in the glycolytic (Embden-Meyerhof) pathway until World War II. One of the then standard biochemistry texts (Thorpe) summarized the position in the 1938 edition as:

> Glycogen → hexose → hexose monophosphate → hexose diphosphate → glycerose phosphate + dihyroxyacetone phosphate → glyceric acid phosphate + glycerol phosphate. Glyceric acid

phosphate → pyruvic acid. Pyruvic acid + glycerol phosphate → lactic acid + glycerose phosphate.

GLYCOGEN BREAKDOWN AND SYNTHESIS

Glycogen Breakdown

Glycogen was isolated by Claude Bernard in 1857 in the course of extensive work on carbohydrate metabolism in animals. When he started his research in 1843 it was supposed that sugar in animals originated exclusively from their food and was combusted in the blood. As a result of his experiments Bernard found the liver was "not an organ for destroying sugar but on the contrary an organ for making it, and I found that all animal blood contains sugar even when they do not eat it." By 1848 Bernard showed in dogs with ligatured blood vessels that if the animal was starved, or had been fed only on meat, there was no sugar in the portal blood, but it was present in blood from the suprahepatic vein, i.e. the liver released glucose, and indeed was the only organ to do so. In 1855 he obtained evidence that glucose could be synthesized in animals from nitrogenous substances, anticipating by nearly 100 years studies on gluconeogenesis. Two years later his ideas on the internal environment were stated in a lecture" . . blood is thus a real environment in which all the tissues liberate the products of their decomposition, and in which they find, for the accomplishment of their functions, invariable conditions of temperature, humidity, oxygenation as well as the nitrogenous, carbohydrate and saline materials without which the organs cannot accomplish their nutrition."

Besides his investigations on glycogen, Bernard was concerned with the differences in carbohydrate metabolism between carnivores and herbivores. In the course of these studies he examined the role of the pancreas in digestion and its ability to hydrolyze starch and fats. For his work on the pancreas Bernard was made a Chevalier of the Legion of Honour, the citation stating that the award was for "excellent work on the musical [vice medical!] properties of the pancreas" (Leicester, 1974).

In a book published in 1878, the year after his death, Bernard first enunciated the concept of homeostasis—a term coined much later (in the 1930s) by Cannon—the essentiality for all living systems to keep constant the conditions of life in the internal environment. Work on the mechanisms for maintaining the constancy of the "mileu interieur" was to be continued by Bernard's associates and lead to modern molecular endocrinology.

Studies on the breakdown and synthesis of glycogen are particularly associated with the work of Carl and Gerti Cori in the 1930s and 1940s. They emigrated to the U.S. from Vienna in 1922 initially to Buffalo, N.Y. and later moved to Washington University School of Medicine at St. Louis in 1931, when they worked together on carbohydrate metabolism, work for which they received a Nobel prize in 1947.

Hydrolysis of starch or glycogen had been known for many years. That glycogen breakdown in muscle occurred by phosphorolysis was shown by Parnas (1937). P_i was essential and hexose monophosphate was produced. Under the conditions employed by Parnas the sugar phosphate was not easily hydrolyzed and was shown to be a mixture of glucose- and fructose-6-phosphates. The Coris (1936) established that the primary intermediate from glycogen phosphorolysis was G-1-P. With minced frog muscle, NAD^+ and most of the P_i was removed by washing; lactic acid could therefore not be formed. The absence of the reducing ability expected from glucose or fructose derivatives and the acid-lability of the phosphorylated product identified it as G-1-P.

Phosphoglucomutase was isolated by the Coris and crystallized by Najjar. Its mechanism of action was suggested from experiments by Leloir in 1951 using [^{32}P]. The enzyme is a phosphoprotein. Leloir showed that the phosphate group was transferred from the enzyme to G-1-P in the course of the reaction to give G-1,6 diP, which then donated the phosphate from its 1-position back to the enzyme, releasing G-6-P:

$$\text{G-1-P} + \text{Enz-}[^{32}\text{P}] \rightarrow \text{G-1,}[^{32}\text{P}]\text{6 diP} + \text{Enz} \rightarrow \text{G-}[^{32}\text{P}]\text{6-P} + \text{Enz-P}$$

Phosphorylase was studied in depth. The enzyme from muscle was different from that catalyzing the same reaction in liver. Muscle phosphorylase but not that from liver, was activated by AMP, an early example of enzyme regulation by a ligand which was not a substrate. [Allosteric regulation was not postulated until the work of Jacob and

Monod (1961)]. Both liver and muscle phosphorylases were rapidly inactivated *in vitro* though some protection was afforded against this by fluoride, suggesting phosphatases might have been active. In 1955 H. Fischer and E.G. Krebs found ATP reactivated aged preparations. Three years later a protein kinase was isolated which phosphorylated phosphorylase to give an active form of the enzyme, phosphorylase a. Reconversion to the inactive b form was also enzymic due to the presence of "PR enzyme," so called because it was originally thought to remove a prosthetic group but was later shown to be a protein phosphatase. Using phosphorylase kinase free from PR enzyme, crystalline muscle phosphorylase b was phosphorylated with $^{32}P(\gamma\text{-}P)ATP$:

$$\underset{\substack{\text{dimer}\\ \text{(monomeric unit, 120 kDa)}}}{\text{2 phosphorylase b}} + 4[[^{32}\text{P}]\text{ATP} \rightarrow \underset{\text{tetramer}}{[^{32}\text{P}]\text{phosphorylase a}} + 4\text{ ADP}$$

^{32}P was incorporated into specific serine residues on the phosphorylase. Similar phosphorylation activated the liver enzyme (Sutherland et al., 1956).

Other experiments anticipating current ideas on the regulation of phosphorylase were reported from Sutherland's laboratory from 1956 onwards. The soluble phosphokinase which phosphorylated liver phosphorylase was activated by a novel heat-stable, dialysable factor formed in the presence of ATP by a particulate fraction from liver. This capacity was enhanced in the presence of glucagon or adrenaline. The factor was identified as adenosine 3′,5′-cyclic phosphate (cyclic-AMP) which had just been isolated by Markham's group in Cambridge during development of chromatographic methods for separating nucleotides. Cyclic-AMP was then shown to activate a protein kinase which had the capacity to phosphorylate a large number of enzymes and other proteins. The kinase (now protein kinase A) was found by Sutherland's group to be widely distributed, being present in brain and glandular tissues (adrenal cortex, thyroid) as well as in liver and muscle.

In *Physiological Reviews for 1960* de W. and M.R. Stetten concluded "[Cyclic-AMP] was possibly of more general importance [than] its role in liver," so percipiently anticipating the Secondary Messenger Hypothesis later proposed by Sutherland.

Glycogen Synthesis

One of the most exciting observations by the Coris (1943), which was confirmed when purified preparations of active phosphorylase a became available, was that the enzyme from plant or animal sources also catalyzed the formation of starch or glycogen, provided traces of polymer were present to act as primer, i.e. to give free 4-OH positions onto which incoming glucose units could be attached. Many generations of students were impressed when they repeated the experiment using potato phosphorylase, and saw the blue color when iodine was added to detect the newly synthesized starch. Glycogen and starch synthesis and breakdown were for a time presumed to be freely reversible, both the forward and backward reactions being effected by phosphorylase a.

There were however some serious objections to this view that glycogen was synthesized by phosphorylase (see Stetten and Stetten, 1960). One argument was kinetic; at pH 7 the equilibrium of the phosphorylase reaction occurs when the ratio of free P_i to G-1-P is 1.3:1. In cells, however, P_i is 100-fold in excess of G-1-P. A second observation showed the forward and backward reactions to require different environments. Baird Hastings and his colleagues (1956-1957) found glycogenesis was favored by high $[K^+]$ and then only if phosphorylase activity was very low. When $[Na^+]$ was high, glycogenolysis always occurred.

A more persuasive, although essentially a negative argument, was that adrenaline, which since the time of Lesser (1920) had been known to promote glycogenolysis in muscle, invariably promoted glycogen breakdown even under conditions when glycogen would have been formed if adrenaline was not present. Further studies (Sutherland, 1951) showed that in the presence of adrenaline G-1-P concentrations rose, as did the amount of phosphorylase a. Adrenaline caused glycogen to break down even though, if synthesis and breakdown were reversible, synthesis should have increased as the concentration of G-1-P rose.

A final argument came from studies of glycogen storage diseases. These are inherited illnesses, often with severely restricted life expectancy, where excessive amounts of glycogen are deposited. The patients are easily fatigued. The first genetic disease to be investigated affecting carbohydrate metabolism was von Gierke's disease (glycogenosis, type I). Here liver glycogen levels of 15% by weight (cf. 5% normal maximum) can be found; the glycogen structure is normal. The Coris (1952) demonstrated the condition was due to a deficiency in glucose-6-phosphatase, the liver enzyme essential for releasing glucose from G-6-P into the blood to maintain blood glucose levels.

McArdle's disease is associated with excessive deposits of glycogen in muscle, and Hers' disease with its deposition in liver. In both cases phosphorylase levels in the affected tissues are very low. In spite of this, glycogen synthesis is unimpaired, which is incompatible with glycogenesis occurring through the action of phosphorylase.

Solution to the problem emerged coincidentally with the appreciation of the difficulties outlined above. Galactose is one of the component monosaccharides in lactose. Since 1908 (von Reuss) it had been known that some children could not tolerate milk. Infants with the condition fail to gain weight after birth, become jaundiced, and may have serious and progressive mental retardation and impaired liver function unless lactose is rapidly removed from their diets. In the 1950s, studies on the utilization of galactose were commenced. Kalckar and Arthur Kornberg isolated enzymes which catalyzed the reaction:

$$\text{UTP} + \text{G-1-P} \leftrightarrow \text{UDPGlu} + \text{PP}$$

Uridine diphosphateglucose (UDPGlu), whose structure was confirmed synthetically by Todd (1954), was detected in all organisms surveyed. It was then found to be essential for the utilization of galactose (Leloir, 1950 et seq.):

$$\text{Gal} + \text{ATP} \rightarrow \text{Gal-1-P}$$

Galactokinase

Gal-1-P + UDPGlu ↔ Glu-1-P + UDPGal

Uridyl transferase

UDPGal ↔ UDPGlu

UDPglucose epimerase

(Inversion of configuration on C4)

Uridyl transferase is the enzyme which is missing in classical galactosemia (Isselbacher, Kalckar, 1956).

Further studies from Leloir's group in Buenos Aires, showed that UDP was also essential in the biosynthesis of sucrose in plants. Its general role in the formation of polysaccharides then emerged. For glycogen, glucose units are added to the 4-OH position on the primer using UDPGlu as carrier, catalyzed by UDP-glycogen synthase (Leloir and Cardini, 1957), a soluble enzyme which they found in liver. In the following year the enzyme was shown to be present in muscle homogenates (Villar-Palasi and Larner, 1958). By 1959 Leloir had concluded that the rate of addition of glucose units to glycogen, catalyzed by UDP-glycogen synthase, was sufficiently fast to account for glycogen synthesis *in vivo* (see Sols, 1961). In 1970 Leloir was awarded a Nobel Prize for his elucidation of the pathways of biosynthesis of polysaccharides and the functions of the uridine diphosphate sugars.

GLYCOLYSIS AND MUSCLE CONTRACTION

Muscle contraction has been studied experimentally since the mid-eighteenth century. In 1745 the Leyden jar was described and it became possible to produce electric discharges in the laboratory. Effects on living tissues began to be explored. Leopoldi Caldoni was one of the first to report (1757) the strong contractions electric shocks produced in skeletal muscle. The experiments of Galvani on the contraction of a frog's leg by an electric current (1780) were more analytical. Galvani recognized the role of the nerve in conducting the electric impulse and that the contractile elements were the muscle. In his notebook for 1835 Schwann described an experimental protocol that would have tested the relation between the length of a muscle contracting in response to an electrical stimulus and the different loads (weights) attached to the muscle. The production of such tension/length diagrams was to become commonplace for electrophysiologists in the next century.

The work of Voit and others (Chapter 3) established the relation between food consumption and energy release. The source of the energy for physical work, its release as heat produced by contracting muscle, and the intracellular basis of that contraction were to become an area for intense activity and controversy for physiologists and biochemists. Of greatest influence were the experiments of A.V.Hill and his associates, mainly with frog sartorius muscle. Contractions of skeletal muscle are of two kinds: isotonic where the muscle shortens but the tension remains constant, and isometric where the muscle maintains its length but the tension increases. Muscles were stimulated to lift weights; heat production was recorded by thermopiles linked to measure the response. By the mid-1920s such experiments had shown that heat was produced in three phases: (1) associated with contraction, (2), maintenance of the contracted state and (3) the recovery phase as the muscle relaxed.

These early experiments were handicapped by the lack of sensitivity of the instruments and their response time. From ca.1940 thermopiles were significantly improved so that they could detect very small increases in temperature in a few msec. Hill was thus able to distinguish an initial phase of heat production, "shortening heat" if the muscle was allowed to shorten, and heat of recovery which might continue for 20-30 minutes. Mathematical expressions were derived (the Hill equation) relating the energy production due to muscle shortening with the load (weight lifted). The work of Fletcher and Hopkins had shown the chemical changes during muscular activity involved the disappearance of muscle glycogen and the equivalent production of lactic acid. Amounts of lactic acid went up as work done by the muscle increased. The consequent fall in pH was rapidly buffered. Oxygen was required during the recovery phase.

Analysis of muscle structure began about the middle of the last century. Microscopists reported a transverse striated appearance when muscle fibers were examined by ordinary light microscopy. With the introduction of the polarizing microscope regions with higher refraction were observed to be birefringent (A bands); those with a lower refractive index were optically isotropic (I bands) (Dobie, 1849). The I bands were bisected by a region of high refractive index extending across the fiber and attached to the sarcolemma (now called the Z band). In the middle of the A band there might be a region of lower refractive index—the H band. The appearance of the fiber, and especially the distinction into "dark" and "light" bands, depended critically on the focusing procedures employed. When the muscle was fixed and stained with the basic dyes used for classical histology (see Chapter 9) contraction was almost inevitably produced. There was therefore much disagreement over the interpretation of the appearance of the muscle and the changes that physiological contraction produced.

The muscle fibrils are embedded in sarcoplasm, each individual fibril showing the banding pattern of the whole fiber (Bowman, 1840). Myosin could be extracted from muscle with strong salt solutions. From the altered appearance of the bands after extraction it was suggested that this protein was a major component of the A bands (Kuhne,1864; Danilewsky, 1881). The localization of myosin in the A bands and of actin in the I bands was convincingly shown by Jean Hanson and Hugh Huxley (1954-1955) in electron micrographs of transected fibers and confirmed after selective extraction to remove myosin (Hasselbach, 1953; Hanson and H.E. Huxley, 1953-1955).

Interference and phase contrast microscopy became available in the 1950s. These obviated the need for fixation and staining (see Chapter 9). It could now be seen that when the fibrils contracted the width of the A bands remained constant with the I bands disappearing as the actin slid between the myosin fibers (A. Huxley and Niedergerke, 1954; H.E. Huxley and Hanson, 1954). Critical examination of the literature (A.F. Huxley,1957) revealed that some elements of this sliding filament theory for muscle contraction had been recognized by earlier microscopists both in the nineteenth and twentieth centuries, particularly the relative constancy of the width of the A bands. When muscle contracts very strongly, however, contractile bands appear. It was probably interpretation of this behavior of highly contracted fibers, with the apparent need for A band shortening, which caused the constancy of the band width in moderate reversible contraction to be overlooked, and the opposing idea of a contractile substance to be advanced.

Results from more chemically oriented experiments seemed to support the contractile substance theory (see Weber, 1958). After extracting muscle with concentrated salt solutions, Weber obtained threads of protein which showed birefringence and appeared to be myosin. Very great excitement was generated when Banga and Szent-Gyorgi (1941-1942) found that in a solution of appropriate K^+ and Mg^{2+} concentrations these threads contractedin the presence of ATP—the first demonstration that ATP might be directly involved in physical work. Straub (1942) showed the protein in the threads was a mixture of myosin with a second protein, actin. The complex of these two proteins, actomyosin, was contractile. Actomyosin threads, however, were not a model for physiologically contracting skeletal muscle, which in reversible contractions gets fatter. Actomyosin threads shortened in the presence of ATP but got thinner because water was extruded (syneresis). Glycerinated muscle fibers, where immersion in glycerol removed ATP, ions, and other soluble factors involved in excitability, retained their capacity to contract physiologically, developed tension, and shortened when ATP was added.

That myosin, a structural protein, also had enzyme activity as an ATPase, had been shown by Engelhardt and Ljubimova (1939-1941). ATP was now found to dissociate actomyosin producing a marked fall in viscosity; the ATP was split to ADP and P_i. Contrasting properties of ATP in muscle systems were also observed. The rigor seen at post-mortem occurred as ATP levels fell. The ATPase activity of myosin could be inhibited by mercurials (which block SH groups on cysteine); with ATPase blocked, ATP caused muscle fibers to relax (Weber and Portzehl, 1952).

The discovery that ATP was not only the source of the energy required for muscle contraction but was apparently directly involved in the contractile process was an enormous stimulus to biochemists and muscle biologists. In the early 1950s attempts were made to determine if the ATP was hydrolyzed to initiate the contraction or was merely involved in the recovery process. Because of the speed with which contraction occurs, experiments had to be performed with amphibian preparations used as cold as possible. First experiments failed to detect either phosphocreatine or ATP breakdown but when frog muscle was treated with iodoacetate to prevent glycolytic regeneration of ATP, phosphocreatine breakdown was proportional to the work done by the muscle (Mommaerts, 1960). Inhibiting creatine phosphokinase with fluorodinitrobenzene allowed ATP breakdown to be demonstrated (Davis, 1962). Rigorous identification of the stages in actin/myosin interactions when changes in adenine nucleotides occur, was not made until the 1970s. It was then realized that resting, but stimulatable ("energized") fibrils had (ADP + P_i) bound onto the myosin catalytic sites, i.e. the ATP was already hydrolyzed in resting muscle. Stimulation causes ADP + P_i to be released, freeing the site on the myosin for interaction with action. Work still continues on the structures of the actin and myosin combining sites, and on the conformational changes in myosin that cause the actin to slide.

The model systems, actomyosin threads, glycerinated fibers, or individual fibrils all contracted. Relaxation was more difficult to achieve *in vitro*. From 1957, Ebashi had obtained granular preparations from muscle from which myofibrils had been removed. The granules, which it was thought might have originated from the sarcoplasmic reticulum, lowered the tension in model systems contracted by ATP. Greater relaxing activity was obtained if the preparations had been treated with oxalate.Suggestions were therefore made that relaxation was linked to the removal of Ca^{2+}. Evidence supporting this came from the parallelism between the ability of chelating agents, such as EDTA and EGTA, to complex with Ca^{2+} and the relaxing effect they had on glycerinated fibers (Ebashi, 1960). The role of Ca ions, their location in the triad system associated with the sarcoplasmic reticulum, and the part played by tropomyosin (Bailey, 1948) and the troponin system both structurally and in the regulation of muscle contraction, were elucidated by Ebashi, Perry, and others in the following years.

REFERENCES*

Axelrod, B. (1960). Glycolysis., Ed.) In Metabolic Pathways Vol. 1, (Greenberg, D.M., Ed.), 2nd ed., Vol. 1, pp. 97-128. Academic Press, New York.

Baldwin, E. (1947). Dynamic Aspects of Biochemistry. Cambridge University Press.

Boyer, P.D. & Segal, H.L. (1954). Sulfhydryl groups and glyceraldehyde-3-phosphate dehydrogenase and acyl-enzyme formation. In: Metabolic Pathways. (Greenberg, D.M., Ed.), 2nd ed., Vol. 1, pp. 520-532. Academic Press, New York.

Dickens, F. (1951). Anaerobic glycolysis, respiration and the Pasteur effect. In: "The Enzymes. (Sumner, J.B. & Myrback, K. Eds.), 1st ed., pp. 624-683. Academic Press, New York.

Hill, A.V. (1926). Muscular Activity. Williams & Wilkins, Baltimore.

Huxley, A.F. (1957). Muscle structure and the theories of contraction. Prog. in Biophy. 7, 255-318.

Leloir, L.F. (1955). The uridine coenzymes. In: Proceedings of the 3rd International Congress of Biochemistry. Liebecq, C., Ed.), pp. 154-162. Vaillant-Carmanne, Liege.

Lundsgaard, E. (1930). Experiments on muscle contraction without lactic acid formation. Biochem. Z. 217, 162-177.

Peters, J.P. & Van Slyke, D.D. (1932). Quantitative Clinical Chemistry. Bailliere, Tindall & Cox, London.

Sols, A. (1961). Carbohydrate metabolism. Annu. Rev. Biochem. 30, 213-238.

Stetten, de W. & Stetten.M.R. (1960.) Glycogen metabolism. Physiol. Rev. 40, 505-537.

Sutherland, E.W. (1952). The effects of epinephrine and the hyperglycemic factor on liver and muscle metabolism *in vitro*. In: Phosphorus Metabolism. (McElroy, W.D. & Glass, B., Eds.), Vol. 2, pp. 577-593. The Johns Hopkins Press, Baltimore.

Szent-Gyorgi, A. (1944). Studies on muscle. Acta Physiol. Scand. 9, suppl. 25.

Thorpe, W.V. (1938). Biochemistry for Medical Students. J.& A.Churchill, London.

Umbreit, W.W., Burris, R.H. & Stauffer, J.F. (1945). Manometric Techniques and Related Methods for the Study of Tissue Metabolism. Burgess Publishing Minneapolis.

Velick, S.F. (1954). The alcohol and glyceraldehyde-3-phosphate dehydrogenases of yeast and mammals. In: The Mechanism of Enzyme Action. (McElroy, W.D. & Glass, B., Eds.), pp. 491-519. The Johns Hopkins Press, Baltimore.

Weber, H.H. (1958). The Motility of Muscles and Cells. Harvard UP.

Wilkie, D.R. (1954). Facts and theories about muscle. Prog. Biophys. 2, 288-324.

Young, M. (1969) The molecular basis of muscle contraction. Annu. Rev. Biochem. 38, 913-950.

*The glycolytic pathway and those for glycogen breakdown and synthesis are shown in Appendix 2.

Chapter 5

ASPECTS OF CARBOHYDRATE OXIDATION, ELECTRON TRANSFER, AND OXIDATIVE PHOSPHORYLATION

MEASUREMENT OF OXYGEN UPTAKE

Studies on cellular respiration were technically more difficult than those on glycolysis because many of the enzymes are mitochondrial and so could not easily be solubilized. The earliest, semi-quantitative procedure for the micro-determination of oxygen utilization was that introduced by Thunberg ca. 1910 known as the Thunberg tube. The tissues which could be used were limited to those soft enough to be chopped or put through a Latapie mincer (Szent-Gyorgi) which gave particles large enough to allow many of the cells to remain intact. After dispersal in a buffer the preparation could be pipetted. Oxidation was followed using nontoxic redox dyes such as methylene blue whose color changed on oxidation or reduction. Many oxidizing systems were able to transfer electrons from substrates such as succinate to redox acceptors.

By the late 1920s quantitative micro-determination of oxygen uptake had been developed in Warburg's laboratory in Berlin based on a manometric technique introduced by Barcroft and Haldane (1902). With this equipment evolution of carbon dioxide or uptake of oxygen could be monitored; Carbon dioxide produced in respiration was absorbed by potassium hydroxide. If bicarbonate buffer was used, acid production caused carbon dioxide to be released. Krebs and others from

Warburg's laboratory were skilled in designing assays which could be adapted to manometric procedures (see Chapter 6). The ease with which measurements could be made by an experienced worker enabled kinetic analyses to be performed rapidly under a variety of conditions. Different ways of preparing tissues were also explored. Fred Waring, an American bandleader, marketed kitchen blenders (Waring blenders) which gave easily handleable suspensions, but intracellular organelles were often extensively damaged so that enzymes were rapidly inactivated. Alternatively the tissues were dispersed using a power-driven, close-fitting pestle, originally made of glass or stainless steel, but now usually of teflon (Potter-Elvejhem homogenizers). Cells were disrupted by the shearing force set up between the rotating pestle and the surrounding walls of the tube. The clearance of the pestle could be selected so that nuclei and mitochondria were relatively undamaged.

Many of the early workers preferred to retain cell organization and diminish organelle damage by using tissue slices (Warburg, 1923). With a soft tissue like liver or kidney, these were cut by hand with a "cut-throat" razor or later, chopped mechanically. Rates of oxygen diffusion into the slices necessarily limited their thickness. With a metabolically active tissue like liver very thin slices were desirable, which were obviously fragile. There were also complications because the outer layer of cells was inevitably damaged.

A very different approach for measuring oxygen uptake now is to use an oxygen electrode with automatic recording. Dissolved oxygen can be reduced electrolytically and this followed continuously by the change in potential. Originally a dropping mercury electrode was used (Vitek, 1933). By 1953 Clark had developed a system for small amounts of material where the electrode compartment was separated from the tissue by a cellophane membrane (now "cling-film" or teflon) across which oxygen diffused rapidly. The method is fast and sensitive and since 1960 has become the method of choice for many respiratory studies.

THE "CYCLE" CONCEPT

The early experiments on respiration in whole animals had established that carbohydrate was completely oxidized to carbon dioxide and water.

Between 1910 and 1930 many dicarboxylic acids were found to be rapidly oxidized; most significantly Batelli and Stern (1910) reported the oxidation of succinate to malate. Fumarate was then shown to be an intermediate and by 1928 the further oxidation of malate to oxaloacetate had been demonstrated. The problem of reconciling the ready oxidation of the 4C dicarboxylic acids with the need to oxidize pyruvate, the 3C product of glycolysis, was ultimately to be solved by Krebs and Johnson (1937). By the early 1920s an initial form of the cycle had emerged from the suggestions of Thunberg, Knoop, and Wieland. This incorporated a condensation of two molecules of acetic acid to give succinate—a reaction suggested by Thunberg but for which no evidence could be produced. In 1930, Toeniessen and Brinkmann reported that muscle perfused with pyruvate gave rise to succinate and formate. They proposed pyruvic acid might undergo a reductive condensation to give the 6C compound 1,4-diketoadipic acid:

$$2\ CH_3COCO_2H \rightarrow \underset{\text{1,4-diketoadipic acid}}{HO_2C{\cdot}CO{\cdot}CH_2{\cdot}CH_2{\cdot}CO{\cdot}CO_2H} + 2H$$

from which succinate and formate could have been derived. The formation of formate by perfused muscle was not substantiated by other workers and was later attributed by Krebs to bacterial contamination which had already bedevilled earlier experiments from Toeniessen's laboratory. The postulated intermediate, 1,4-diketoadipic acid, was synthesized and found to be oxidized extremely slowly, so reinforcing the conclusion that this cycle was untenable.

The first suggestion that substrates in carbohydrate oxidation might exert catalytic effects on the oxidation of other intermediates (cf.earlier demonstration of such action in the urea cycle by Krebs and Henseleit, 1932; see Chapter 6) arose from the work of Szent-Gyorgi (1936). He demonstrated that succinate and its 4C oxidation products catalytically stimulated the rate of respiration by muscle tissues. He also observed that reactions between the 4C intermediates were reversible and that if muscle was incubated with oxaloacetate, fumarate and malate made up 50-75% of the products, 2-oxoglutarate 10-25% and, significantly, 1-2% of the C was converted to citrate. These observations were confirmed the same year by Baumann and Stare, so that Szent-Gyorgi could conclude that respiration involved the oxidation of a 3C sugar derivative (triose) via oxaloacetic acid, thus linking glycolysis to the oxidation of the dicarboxylic acids. The precise details of the linkage remained speculative.

Citric acid was isolated by Scheele from lemons in 1784 and was soon found to be widely present in plants. In 1911 Thunberg showed that the addition of citrate increased respiration in muscle homogenates. The importance of citric acid (6C) in carbohydrate oxidation was demonstrated by H.A. Krebs and Johnson, his research student, in 1937. Krebs received his training in research methods in Warburg's Institute in Berlin before taking up a clinical post in Freiburg where he formulated the urea cycle (see Chapter 6). Krebs left Germany for England in 1933, and moved to Sheffield from Cambridge in 1935. Before he left Germany he had been reviewing the Thunberg-Knoop-Wieland cycle and had begun work on acetate metabolism in guinea-pig and rat liver slices (see Holmes, 1991).

When his studies on carbohydrate oxidation restarted in Sheffield, Krebs' experiments included studies on the anaerobic dismutation of pyruvate by bacteria and various animal tissues. Assuming the role for the dicarboxylic acids postulated by Szent-Gyorgi, the main question was the route by which the carbon atoms of pyruvate were converted to succinate. In May 1936 Krebs had observed that if 2-oxoglutarate was added to pyruvate, the yield of succinate was enormously increased. In his notebook written that year (Holmes, 1993) Krebs postulated:

[either] malic acid + acetic acid → citric acid -2H

or malic acid + pyruvic acid = [C7 tricarboxylic acid]

→ citric acid + $2CO_2$

By the beginning of October that year results from Johnson's experiments allowed Krebs to report at a Biochemical Society meeting in Cambridge: "If pyruvic acid is added to tissues under anaerobic conditions, together with malic acid or oxaloacetic acid, very considerable quantities of citric acid are formed."

Besides Szent-Gyorgi and Krebs, other groups were attacking the problem of carbohydrate oxidation. Weil-Malherbe suggested: "It is probable that the further oxidation of succinic acids passes through the stages of fumaric, malic, and oxaloacetic acid; pyruvic acid is formed by the decarboxylation of the latter and the oxidative cycle starts again." K.A.C. Elliott, from the Cancer Research Laboratories at the University of Pennsylvania, also proposed a cycle via some 6C acid.

In February the next year Martius, then working in Knoop's laboratory in Tubingen, demonstrated that the oxidation product from citrate was 2-oxoglutarate, a 4C compound. It seems likely (Holmes, 1993) that Krebs and Johnson were thus prompted to test the effects of citrate on the respiration of pigeon-breast muscle. Respiration with this preparation declined sharply after 20-40 minutes. If citrate was added the respiration was stimulated and its fall-off delayed. Malonate was known from the work of Quastel (1924-1931) to block the oxidation of succinate by succinic dehydrogenase. When malonate was added with citrate to a respiring pigeon-breast muscle preparation, succinate accumulated, confirming that a 6C compound was metabolized to a 4C derivative.

In June, 1937 Krebs sent a letter to *Nature* entitled "The Role of Citric Acid in Intermediate Metabolism of Animal Tissues." It stated: "Triose reacts with oxaloacetic acid to form citric acid and in the further course of the cycle oxaloacetic acid is regenerated. The net effect of the cycle is the complete oxidation of triose." Because he had received a surplus of letters, the editor wrote to Krebs that he would be unable to publish the communication for 7 to 8 weeks. The paper with some modifications and additions was therefore submitted to, and accepted by, *Enzymologia.*

In the next 2 to 3 years further experiments, particularly by Eggleston, who had joined Krebs in January 1936, confirmed and extended the observations. Careful quantitative evaluation of the data indicated that citrate like fumarate (Szent-Gyorgi) and like ornithine in the urea cycle exerted a catalytic effect on muscle metabolism. If arsenite, which blocks 2-oxoglutarate oxidation, was added with citrate to a respiring pigeon-muscle preparation, 2-oxoglutarate accumulated.

A final demonstration was that 4C intermediates gave rise to citric acid. This was shown by Krebs with oxaloacetate as substrate under anaerobic conditions so that oxidation was blocked and citrate levels built up.

The tricarboxylic acid cycle was therefore validated, having been tested not only in pigeon-breast muscle but also with brain, testis, liver, and kidney. The nature of the carbohydrate fragment entering the cycle was still uncertain. The possibility that pyruvate and oxaloacetate condensed to give a 7C derivative which would be decarboxylated to citrate, was dismissed partly because the postulated compound was oxidized at a very low rate. Further, work on the oxidation of fatty acids (see Chapter 7) had already established that a 2C fragment like acetate was produced by fatty acid oxidation, en route for carbon dioxide and water. It therefore seemed likely that a similar 2C compound might arise by decarboxylation of pyruvate, and thus condense with oxaloacetate. For some considerable time articles and textbooks referred to this unknown 2C compound as "active acetate."

When in later years Krebs reviewed the major points which had to be established if the cycle was to be shown to be operative in cells, the obvious needs were to find the presence of the required enzymes and to detect their substrates. As the substrates are present in the cycle in catalytic amounts their accumulation required the use of inhibitors. Krebs also stressed that rates of oxidation of the individual substrates must be at least as fast as the established rates of oxygen uptake *in vivo,* an argument first used by Slator (1907) with reference to fermentation: "A postulated intermediate must be fermented at least as rapidly as glucose is." (See Holmes, 1991). This requirement did not always appear to be met. In the early 1950s there were reports that acetate was oxidized by fresh yeast appreciably more slowly than the overall rate of yeast respiration. It was soon observed that if acetone-dried or freeze-dried yeasts were used in place of fresh yeast, rates of acetate oxidation were increased more than enough to meet the criterion. Acetate could not penetrate fresh yeast cell walls sufficiently rapidly to maintain maximum rates of respiration. If the cell walls were disrupted by drying this limitation was overcome, i.e. if rates of reaction are to be considered, it is essential to ensure substrates have adequate access to their enzymes.

SOME STEPS IN THE TRICARBOXYLIC ACID CYCLE

Oxidative Decarboxylation of Pyruvate

The Intracellular Function of Vitamin B_1

Elucidation of the intracellular function of vitamin B_1 was largely the result of studies by R.A. Peters and his associates in Oxford in the 1930s. Peters had been encouraged to work on water-soluble vitamins during his period in Cambridge with Gowland Hopkins. In 1929 high levels of lactate were detected in blood from Japanese patients suffering from beri-beri. This could be explained if pyruvate oxidation was blocked. Two years later another observation of lactate accumulation was made with pigeons which were being fed polished rice (R.B. Fisher, 1931). The birds soon adopted a characteristic head retraction (opisthotonus) and were unable to fly.

Peters was convinced that the disorder in the pigeons was central in origin. He therefore began an exhaustive study of intracellular pyruvate metabolism using brains from normal and B_1-deficient pigeons. Pigeon brain was disintegrated by gentle chopping with a bone spatula. Connective tissue was removed by filtration through muslin. If glucose or lactate were used as substrates this "brei" showed depressed oxygen uptake when prepared from B1-deficient birds. With succinate, oxygen uptake was normal. In a dramatic experiment Peters was able to show that injecting thiamine into the subarachnoid space reversed the neurological signs shown by the pigeons. This method of administering the vitamin was necessary in order to eliminate difficulties the vitamin would otherwise have had in traversing the blood-brain barrier. Forty minutes after injection the bird was again able to fly and the capacity of the brain to oxidize lactate was also restored.

In 1936 Peters introduced the concept of a "biochemical lesion" as a basis of human disease, an illness attributable to an identifiable biochemical disorder. This idea was very influential in directing the attention of clinicians to the importance of biochemistry as a discipline in the training of medical students. Besides the example of vitamin B_1, other instances of specific biochemical lesions were soon identified in genetic diseases arising from enzyme deficiencies as in alcaptonuria.

The way in which thiamine participated in the oxidation of pyruvate became clearer when Lohmann and Schuster (1937) showed vitamin B_1 to be present intracellularly as thiamine pyrophosphate. In yeast, decarboxylation of pyruvate yielded ethanal which was reduced by alcohol dehydrogenase to give ethanol. A cofactor was needed for this decarboxylation, co-carboxylase. Like the cofactor needed in animal cells for the decarboxylation of pyruvate, cocarboxylase was found to be identical to thiamine pyrophosphate. Vitamin B_1 thus became the first vitamin whose intracellular function as a coenzyme had been established *in vitro*. Another aphorism therefore arose about vitamins—B vitamins are (parts of) coenzymes—an idea that was to be completely confirmed.

Insight into the mechanism of action of vitamin B_1 in decarboxylation came later from Breslow (1957) who used deuterium labeling to show that the H on position 2 in the thiazole ring was dissociable. The pyruvate is attached there prior to its decarboxylation. In yeast the 2C fragment is then released as ethanal and subsequently reduced by alcohol dehydrogenase to give ethanol. In animals the fragment is further oxidized to give acetyl coenzyme A, i.e. "active acetate."

Heavy Metal Poisoning: The Involvement of Lipoic Acid

Heavy metals and their derivatives have been used from time immemorial for cosmetic (e.g. white arsenic and white lead preparations) and therapeutic (e.g. mercury in the treatment of syphilis) purposes. Excessive or prolonged use caused premature death. Signs and symptoms of lead poisoning were first described by Hippocrates, and arsenic in various forms was used historically as a poison (see D.L. Sayers, *Strong Poison*, 1930). Acute poisoning causes severe gastrointestinal disturbances. Chronic low doses, arising in various parts of the world by continually drinking water with particularly high arsenic levels, produces hyperpigmentation in the skin and is said to improve the sleekness of human hair and horses' coats.

The use of inhibitors which react with thiol (SH) groups in proteins were mentioned earlier. Arsenic (Ehrlich, 1909) and mercury are also SH inhibitors. Arsenic was employed experimentally either as arsenite (AsO_2^-) or as phenyl arsenoxide, ($C_6H_5 \cdot AsO$). Towards the end of World War I. a very toxic arsenic derivative, lewisite, ($ClCH{=}CH \cdot AsCl_2$), known as the "dew of death," was synthesized by Lewis but was not used in gas warfare. In the late 1930s as World War II became increasingly probable, research was urgently directed to the development of antidotes against lewisite poisoning. The Oxford workers noted that arsenite prevented the stimulation of oxygen uptake produced by the addition of thiamine to B_1-deficient pigeon-brain brei. Lewisite had a similar inhibitory effect on pyruvate oxidation, the basis of its acute toxicity. Attempts were then made to reverse the inhibition. The monothiol thioglycollate, ($HS \cdot CH_2 \cdot CO_2^-$), had some protective action. Kerateine obtained by treating hair with cyanide was shown to bind arsenic through two thiol linkages. It was therefore suggested that a dithiol which could form a 5- or 6-membered ring, might be more effective than a monothiol in reversing arsenic and lewisite inhibition. The compound, 1,2-dimercaptopropanol [British Anti-Lewisite, (BAL)] was synthesized by Stocken and Thompson (1940) in the expectation that it would be absorbed through the skin and reach intracellular sites rapidly. BAL rapidly reversed the toxic effects of lewisite and mercury poisoning.

Following from this the suggestion arose that a dithiol might be an essential component in oxidative decarboxylation. This was established by Reed and Gunsalus and their colleagues in 1952 when the two groups identified a new growth factor required by microorganisms for the oxidation of pyruvate. The factor was 6,8-dithioctanoic acid (lipoic acid). Microbial reactions needing lipoic acid were inhibited by arsenite, and this was reversed by BAL. The requirement for lipoic acid in oxidative decarboxylation reactions in animals was then rapidly demonstrated, acetyl lipoamide serving as the donor of the acetyl group to give acetyl coenzyme A.

Acetyl Coenzyme A (Acetyl CoA)

By the late 1930s it was widely accepted that "active acetate" arose from pyruvate decarboxylation and fatty acid oxidation. Acetate itself was relatively inactive. In 1939 Lipmann had moved to the United States, initially as a Research Associate at Cornell Medical School, then via Massachusetts General Hospital to Harvard and finally, in 1957, to the Rockefeller Institute. Lipmann identified the start of his successful studies on biological acetylation (for which he was awarded a Nobel prize in 1953, the same year as the award to Hans Krebs) to his observation that some microorganisms oxidized pyruvate to give a very reactive and unstable acetate derivative, acetyl phosphate. This compound was a "high-energy phosphate compound" (see below). It was not however oxidized by brain brei (Ochoa, Peters, and Stocken, 1939) and so could not be an intermediate in the oxidation of pyruvate in animals. Its properties, and especially its extreme lability, also did not appear consistent with the properties of acetylating systems in animals.

Two of these systems were studied as models—the acetylation of choline in brain to give acetyl choline (Hebb, Nachmansohn), and of sulfanilamide (the active component in prontosil, Chapter 3) in liver (Lipmann). Sulfanilamide is rapidly inactivated by acetylation on the *p*-amino group and then excreted. Sulfanilamide is easily diazotized; the diazonium salt formed can be coupled with *N*-(1-naphthyl)ethylenediamine dihydrochloride to give a pink derivative (Bratton and Marshall, 1939). This formed the basis for an elegant colorimetric assay. Only the free *p*-amino group reacts, so that as acetylation proceeded color formation diminished.

ATP and magnesium were required for the activation of acetate. Acetylations were inhibited by mercuric chloride suggesting an SH group was involved in the reaction either on the enzyme or, like lipoic acid, as a cofactor. Experiments from Lipmann's laboratory then demonstrated that a relatively heat-stable coenzyme was needed—a coenzyme for acetylation—coenzyme A (1945). The thiol-dependence appeared to be associated with the coenzyme. There was also a strong correlation between active coenzyme preparations and the presence in them of pantothenic acid—a widely distributed molecule which was a growth factor for some microorganisms and which, by 1942-1943, had been shown to be required for the oxidation of pyruvate.

The active form of acetate, acetyl CoA, was finally isolated by Lynen and Reichert in 1951 following studies of fatty acid oxidation (Chapter

6). Work on fatty acid oxidation and other reactions then showed CoA was required as an acceptor not just for acetate but for many acylation reactions to give acyl CoA, (R·CO-CoA).

Problems with Citrate

Asymmetric Syntheses

The details of the oxidation of citrate to 2-oxoglutarate, via *cis*-aconitic acid (Martius and Knoop, 1937) and isomerization to yield isocitrate (Wagner-Jauregg and Rauen, 1935) had been determined by the time the citric acid cycle was formulated. The order of the 6C intermediates was less certain. Evans and Slotin (1941) and Wood, Werkman, and their colleagues (1942) found carbon dioxide fixation occurred in animal tissues, particularly liver. This allowed isotopically labeled oxaloacetate to be synthesized from pyruvate and $H^*CO_3^-$ with the labeled $^*C$ atom in an unequivocal position. With uniquely labeled oxaloacetate it was possible to examine the behavior of citrate and isocitrate by analyzing isotope distribution in an end-product of their metabolism, 2-oxoglutarate, which could be trapped as the phenylhydrazone. As citric acid is a symmetrical molecule, classical theoretical chemistry stated the isotopic label should be equally distributed between the two positions: $-CH_2{}^*CO_2^-$ and $-CO^*CO_2^-$ in the oxoglutarate. When the distribution was analyzed however, all the radioactivity was in the carboxyl group in position 1, $-CO^*CO_2^-$, the carboxyl group adjacent to the α-keto group.

The consequent interpretation, accepted by Krebs in his review of the tricarboxylic acid cycle in 1943, was therefore that citric acid could not be an intermediate on the main path of the cycle, and that the product of the condensation between oxaloacetate and acetyl CoA would have to be isocitrate, which is asymmetric. This view prevailed between 1941 and 1948 when Ogston made the important suggestion that the embarrassment of the asymmetric treatment of citrate could be avoided if the acid was metabolized asymmetrically by the relevant enzymes, citrate synthase and aconitase. If the substrate was in contact with its enzyme at three or more positions a chiral center could be introduced.

This interpretation offered by Ogston was confirmed by further isotope experiments (Potter and Heidelberger, 1949; Martius and Schorre, 1950). Ochoa and his colleagues also demonstrated that in pigeon-liver preparations which had been freed from aconitase, thus preventing isomerization between citrate and isocitrate, citrate was the product of the condensation of oxaloacetate and acetate.

Lethal Synthesis

Another feature associated with the metabolism of citrate emerged from further work from Peters' laboratory. Some South African plants (Gibflaar) (Dichapetalum cymosum) eaten by cattle, were very toxic. Fluoracetate, (FAc, or $CH_2F{\cdot}CO_2^-$) was isolated from the plant preparations (Marais, 1943) and shown to be lethal in animals in particularly low concentrations. Initially it was supposed that the toxicity might be due to the compound behaving analogously to the effects of F^- on Mg-dependent enzymes like enolase. Alternatively FAc might have acted as an alkylating agent like bromacetate or iodoacetate, and blocked thiol enzymes such as glyceraldehyde-phosphate dehydrogenase. When tested, FAc showed neither of these types of inhibition since the C-F bond is very stable.

The key to unraveling the toxicity of fluoracetate came from observations of Buffa and Peters (1949) that in animals treated with FAc, considerable quantities of citrate accumulated in some tissues. Oxygen uptake was also diminished. The citric acid cycle was thus implicated as the site of inhibition. Fluorcitrate was then isolated from the affected tissues. It was found to be a powerful competitive inhibitor of aconitase, thus blocking citrate oxidation. The suggestion was therefore made that fluoracetate was toxic not in itself, but because it was metabolized in cells via fluoracetyl CoA to give a toxic derivative, an example of "lethal synthesis"—the capacity of organisms to metabolize nontoxic compounds and convert them to potentially lethal products.

TERMINAL OXIDATION: THE CYTOCHROME CHAIN

The mechanism by which oxygen became involved in respiration was discovered in outline between 1920 and 1939 although important details, especially of the link between reduced pyridine nucleotides and the cytochromes, were not unraveled until the 1950s. The main thrust of the work came from Otto Warburg and his associates, initially at the Kaiser Wilhelm Institute for Biology in Berlin-Dahlem, and, between 1931 and his death in 1970, from Warburg's Kaiser-Wilhelm Institute for Cell Physiology. Warburg had been a student of Emil Fischer before going to the Radiation Institute at Berlin-Charlottenberg where Helmholtz and Max Planck had worked. The "family tree" (H.A. Krebs, 1981) was continued as many of those who contributed significantly to biochemistry between 1930 and 1960 spent time in Warburg's laboratory before World War II.

The Cytochromes

Simultaneously with the elucidation of the tricarboxylic acid cycle was the discovery of the means by which oxygen was utilized in cells for terminal oxidation. An abortive start on this had been made by MacMunn (1886), a medical practitioner and the author of a major text on spectroscopy. In a paper for the Royal Society (1886) MacMunn had concluded:

> Thus...throughout the animal kingdom we find in various tissues a class of pigments whose spectra show a remarkable resemblance to each other; they are allied to the hemochromogens, the bands of which are closely imitated by the histohematins ...Their bands are intensified by reducing agents and enfeebled by oxidizing agents; they accordingly appear to be capable of oxidation and reduction and are therefore respiratory ... These observations appear to me to point out the fact that the formation of carbon dioxide and the absorption of oxygen takes place in the tissues themselves and not in the blood.

The work was strongly criticized by Hoppe-Seyler and others. They thought the pigments described by MacMunn were artifacts arising from the breakdown of hemoglobin, in spite of MacMunn's use of perfusion to remove blood and his reiterations that the main absorption bands of his pigments differed from those of known derivatives of myo- or hemoglobin. Further, the bands were present in insects and yeast which do not have hemoglobin. Because of the weight of opinion against his interpretation, MacMunn's discovery of the cytochromes was ignored for almost 40 years. He died in 1911, 14 years before the rediscovery of his work by Keilin.

Investigations however continued on the properties of hemoglobin and its constituents. The porphyrin nucleus had been identified in the urines of patients suffering from porhyria, a disease in which, due to errors in the synthesis of heme, a wine-red urine is voided. The correct structure for the porphyrin ring was proposed by Kuster in 1913 and confirmed by synthesis by H. Fischer in 1929. By this time too it was known that the products formed when hemoglobin was acidified or made alkaline were hemin, ($Hb{\cdot}Fe^{3}$ Cl), or hematin, ($Hb{\cdot}Fe^{3+}OH$), both containing ferric iron. Oxygen carriage was a function of hemoglobin which contained ferrous iron. When hemoglobin was oxidized without protein denaturation, methemoglobin $Hb{\cdot}Fe^{3+}$ was formed, (Kuster, 1910), which was not an oxygen carrier, i.e. the distinction between oxygenated and oxidized hemoglobin had been identified, although it was not fully appreciated before the studies of Conant and R. Hill (1925-1927).

Attention next turned to intracellular reactions, specifically to dehydrogenation—the removal of hydrogen from substrates, especially tricarboxylic acid cycle intermediates. Succinate dehydrogenase had been identified by Thunberg and others. Thunberg postulated that in oxidation, oxygen did not react directly with the substrate, but with hydrogen, to produce water. In 1932 Wieland remarked on the similarities between enzymic dehydrogenation, with hydrogen being accepted by e.g: methylene blue, and non-biological hydrogenation where hydrogen was introduced into acceptors using finely divided metals like palladium as catalysts.

In contrast, Warburg focused attention on iron-catalyzed biological oxidations. He observed (1924) that oxygen uptake in particulate extracts from sea urchin eggs was promoted by the addition of iron, a unique reaction of urchin eggs which have stores of the porphyrin nucleus into which iron can be inserted to give heme pigments. Looking at sources of iron in biological systems, he found that if hemin was incinerated to give "blood charcoal" it was a very effective catalyst for oxidation reactions, (e.g., leucine, being oxidized to valeric aldehyde, ammonia, and carbon dioxide. Charcoal prepared from pure sucrose did not possess the same catalytic activity. The importance of iron was confirmed from the inhibition produced by hydrogen sulfite, carbon monoxide, and cyanide, all of which combined with Fe^{2+} or Fe^{3+} ions and blocked the oxidations.

The lethal effects of carbon monoxide on hemoglobin had been analyzed by Claude Bernard and shown to be due to the formation of an iron-carbonyl compound. In 1891 Mond and Langer showed that iron pentacarbonyl could be dissociated by light, and in 1897 J.S. Haldane and J.L. Smith found light would decompose the carbonyl compound of hemoglobin. Other metal carbonyls are not photodecomposed.

When visiting Berlin, A.V. Hill drew Warburg's attention to the photoreversibility of iron-carbonyl formation reported by Haldane. Warburg made very effective use of this photosensitivity of iron-carbonyls to analyze the mechanism of oxygen activation in cells. Model studies by Krebs, who was then working in Warburg's laboratory, demonstrated that catalysis of oxygen uptake by nicotine-heme or pyridine-heme was inhibited by carbon monoxide. If light was admitted the inhibition was lessened. When a biological system, yeast, was used, photoreversibility was again found. Warburg next examined the "action spectrum" for reversibility in carbon monoxide-treated cells. Using monochromatic light he determined the effectiveness with which light of different wavelengths reversed the respiratory inhibition. The spectrum found was in excellent agreement with the spectrum of the hemin carbonyl, strongly implicating an iron-heme compound as a catalyst ("Atmungsferment") in cellular oxidations.

From his experiments Warburg emphasized the importance of oxygen activation in biological oxidations, while Wieland vigorously supported hydrogen activation, opposing views which Gowland Hopkins at the Physiological Congress in Stockholm in 1926 endeavored to reconcile, with the observation: "[These theories] are mutually incompatible only when either is expressed in too dogmatic form." (See Florkin, 1975). This sentiment could well have been applied to other biochemical controversies both past and to follow.

The link between the dehydrogenation of tricarboxylic acid cycle substrates and oxygen uptake was clarified by Keilin between 1925 and

1939. When working in Paris before World War I. Keilin had been an entomologist. It was natural therefore for him to use insects when, after moving to Cambridge in 1915, he began to study respiration and biological oxidations. Keilin looked at very thin, transparent tissue preparations with a reversion spectroscope. Absorption bands in the visible region of the spectrum could be seen and their position identified by reference to the known position of the sodium bands in sunlight (Fraunhofer lines). In complete confirmation of MacMunn's original observations, and at the time in ignorance of them, Keilin found there were absorption bands in yeast preparations and in vertebrate muscle which had been perfused to wash out blood. The intensity of the bands was greatly increased in the absence of oxygen, with cyanide, or if the muscle was exercised. Particular weight was attached to experiments with the common wax moth, which does not have hemoglobin, and could be studied while still alive, although immobilized. Here the absorption bands and their changes in intensity could be seen in normal living material.

The use of the reversion spectroscope enabled the position of the absorption bands to be determined accurately and to be conclusively distinguished from hemoglobin and myoglobin. It became clear that there were three different intracellular respiratory catalysts—cytochromes *a,b,c*—common to animals, bacteria, yeast and higher plants. In 1925 a preliminary scheme for the passage of O_2 from blood to tissue was proposed:

substrate ← cytochromes ← muscle Hb ← blood Hb ← O_2
(myoglobin)

By 1929 the scheme had been extended to include the dehydrogenases:

$+O_2$
H → leucoindophenol → indophenol
↗
Dehydr(ogen)ase→ substrate → cyt *a*,(*b*), *c* ← O← oxidase← O_2
↘
H → cyt *b*

Cytochrome *b* was distinguished from cytochromes *a* and *c* because it appeared to be autoxidizable in the presence of cyanide.

Cytochrome *c* can easily be extracted from tissue particles by dilute salt solutions. It was isolated by Keilin and Hartree in 1930 and shown to contain a porphyrin ring structure. In 1933 Zeilen and Reuter established that cytochrome *c* was a heme (iron-porphyrin) protein. Slightly different forms of cytochrome *a* were distinguished in yeast and bacteria by Keilin in 1934 and the different properties of cytochrome *a* and a_3 by Tamiya et al. in 1937. The identity of cytochrome a_3, the enzyme which activates oxygen with Warburg's atmungsferment, was proposed by Keilin in 1939. Cytochrome a/a_3 was renamed cytochrome oxidase by Malcolm Dixon (1939). The oxidation route then offered was:

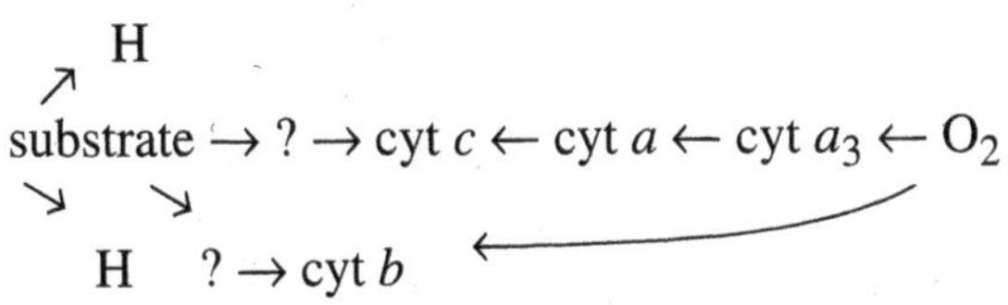

The order in which the heme compounds reacted with oxygen was not firmly established nor was the distinction in function between cytochrome *c*, an electron carrier, and cytochrome oxidase, the enzyme catalyzing the oxidation of ferrocytochrome *c* by oxygen to give ferricytochrome *c*, (cyt $c{\cdot}Fe^{3+}$).

Redox potentials were also used to arrange the electron carriers in their correct order. This procedure was applied to the cytochromes by Coolidge (1932). There were however serious difficulties. Electrochemical theory applies to substances in solution: the values obtained are significantly affected by pH and the concentrations of the different components. Of the members of the electron transport chain only the substrates NAD^+, $NADP^+$, and cytochrome *c* are soluble. The other components were difficult to extract from tissue particles without altering their properties. Further, it was hard to determine their concentration and to decide on appropriate values for pH and oxygen concentration. Nevertheless, mainly from work by Ball (1938), at the time in Warburg's laboratory, an approximate order of redox potentials was drawn up:

substrate	< NAD^+, $NADP^+$	< cyt *b*	< cyt *c*	< cyt oxidase	< O_2
Redox potential, in volts	-0.3	0	+0.25	?	+0.8

By 1949 low temperature spectroscopy had been introduced. With this technique Keilin and Hartree detected a further component in the electron transfer chain which had a sharp band at 552 nm. They later showed it to be identical with cytochrome c_1, which had first been observed by Yakushiji and Okunuki (1940) during succinate oxidation by cyanide-inhibited beef heart muscle. As the oxidation of cytochrome c_1 was accelerated by cytochrome *c*, Okunuki and Yakushiji (1941) had placed c_1 in the chain in the order

$$b - c_1 - c - a - a_3$$

This was confirmed by Keilin and Hartree using antimycin A as an inhibitor. The antibiotic blocked the reduction of cytochrome c_1 by NADH or succinate but did not block the reduction of cytochrome *b*. This site-specific inhibition brought antimycin A into popular use by biochemists in the analysis of electron transfer and oxidative phosphorylation.

Flavoproteins

By 1939 the outstanding problem was to link the dehydrogenases acting on the substrates of the tricarboxylic acid cycle with the electron transport chain. In practice this meant finding how the transfer of hydrogen from NADH or NADPH could be tied to the passage of electrons from cytochrome *c* to a/a_3 and oxygen. The importance of flavoproteins started to emerge in the 1930s. In 1932 Szent-Gyorgi obtained a yellow preparation from heart muscle whose color disappeared on reduction and reappeared when oxygen was readmitted. Szent-Gyorgi therefore suggested the affected compound might be involved in respiration. In the same year Warburg and Christian obtained a "Gelbferment" (yellow enzyme) from the Lebedev yeast extract. The material contained a yellow prosthetic group which was reduced by the yeast and reoxidized by oxygen. When the "ferment" was dialyzed against dilute acid the prosthetic group was dissociated (Theorell, 1934-1935) and shown to contain the B vitamin, riboflavin, and phosphate, as the mononucleotide (FMN). Other flavoproteins were then identified, notably D-amino acid oxidase (see Chapter 6) (Warburg and Christian, 1938) which contained flavin adenine dinucleotide, (FAD), as prosthetic group. Succinic dehydrogenase was shown by F.G. Fischer (1937-1939) to be a flavoprotein.

Redox potentials of the flavoproteins were reported but the values were significantly altered if the prosthetic group was dissociated from its apoprotein, so rendering the ordering of the flavoproteins by their potentials very difficult. It did, however, become clear that there were different classes of flavoproteins: those like D-amino acid oxidase which catalyzed reactions in which molecular oxygen participates directly (flavoprotein oxidases), and those in which oxygen was not itself a reactant and which often, though not invariably (succinic dehydrogenase) also involved NADH or NADPH (G-6-PDH; Warburg and Christian).

By this time it was known, mainly from Warburg's laboratory, that NAD^+ was the main hydrogen acceptor for dehydrogenase reactions associated with glycolysis and the tricarboxylic acid cycle. The need, therefore, was to find a system linking NADH to the cytochromes which reacted sufficiently rapidly to account for the known rate of oxygen uptake. D.E. Green's and von Euler's groups reported such an enzyme in 1938. It was widely distributed and was a flavoprotein. Their preparations were then found to be mixtures but did contain an enzyme which rapidly reoxidized NADPH, transferring electrons to cytochrome *c* (cytochrome *c* reductase; Haas et al., 1940). A system to reoxidize NADH and reduce cytchrome *c* remained elusive although flavoproteins seemed to be involved.

The Electron Transfer Chain Post 1945

Between 1945 and 1960 the links between succinate, its dehydrogenase, NADH, and the cytochrome chain were aggressively re-examined by Slater, Chance, and David Green's groups. Slater used the well-established inhibitor approach (CO, CN^-, azide to block cytochrome oxidase, and BAL, antimycin A, and amytal to stop electron transfer from succinate or NADH to cytochrome *c*) to show that electron transfer from the succinate and NADH pathways converged, and so competed for the final components of the pathway:

$$
\begin{array}{llcl}
 & \text{succinate} \rightarrow \text{dehydrogenase} \rightarrow \text{cyt } b & \searrow & \\
\text{NADH} \rightarrow & \text{dehydrogenase} & \longrightarrow & \text{factor} \rightarrow \text{cyt } c_1 \\
 & \text{(diaphorase)} & & \text{(Slater's factor)} \\
 & \rightarrow \text{cyt } c \rightarrow \text{cyt } a \rightarrow \text{cyt } a_3 \rightarrow O_2 & &
\end{array}
$$

Reviewing the criteria for inclusion of components into the electron transport chain, Slater (1958) highlighted considerations previously advanced by H.A. Krebs as necessary to establish a pathway, namely that the amounts of enzyme present must be commensurate with enzymic activity in the preparation, activity should be fully restored by the reintroduction of the postulated component into an inhibited or depleted preparation, and that the rates of oxidation and reduction of components must be at least as great as those in the system overall. Reduction of cytochrome *b* by the systems then in use was thought by Chance (1952) and Slater (1958) to be too slow for the inclusion of this cytochrome into the main chain.

By 1950, Hogeboom and Schneider had prepared relatively pure mitochondria (Chapter 9) which could be used as the oxidizing system rather than crude particulate preparations like the heart muscle system of Keilin and Hartree. The problem still remained for the extraction and identification of the components from the (inner) mitochondrial membrane without denaturation.

A further important technical innovation was introduced in the early 1950s. The earlier respiring preparations and mitochondria provided turbid suspensions which could not be used for simple spectrophotometry because light scattering by the particles invalidated absorption measurements. Mainly as a result of the work of Chance and G.R. Williams, double-beam spectrophotometers were developed from 1950 onwards. With these, monochromatic light passing through the mitochondrial preparation could be compared directly to that transmitted through mitochondria in the experimental system. Light scattering is almost independent of wavelength in the visible region.

Once the control and experimental light transmissions were balanced at an appropriate wavelength, absorbing in the experimental system could be monitored at a different wavelength suitable for the component under investigation. Measurements were thus made in intact, actively respiring organelles.

Using double-beam spectrophotometry, Chance and Williams (1955) were able to follow changes in the absorption of individual components of the electron transfer chain in mitochondria directly. Furthermore, their mitochondria, but not preparations made by the Keilin-Hartree procedure, exhibited oxidative phosphorylation. Indeed oxidation of the substrate and, indirectly, the formation of ATP (see below) could be followed in the same cuvette. The preparations were initially anaerobic, giving intense absorption bands. When oxygen was admitted, the changes in absorption could be measured at the wavelength uniquely characteristic of the particular reactant. This led to confirmation of Keilin's original scheme, and the reintroduction of cytochrome *b* into the direct pathway as well as the identification of new components. It was also possible to determine amounts of the individual constituents and their normal redox state. In mitochondria, in the presence of oxygen and with succinate as substrate, cytochrome a_3 is 87% oxidized, cytochrome *a* 69% oxidized, cytochrome *c* 64% oxidized, and NADH 100% reduced.

The first of these new, electron transferring components was coenzyme Q (CoQ). Festenstein in R.A. Morton's laboratory in Liverpool had isolated crude preparations from intestinal mucosa in 1955. Purer material was obtained the next year from rat liver by Morton. The material was lipid soluble, widely distributed, and had the properties of a quinone and so was initially called ubiquinone. Its function was unclear. At the same time Crane, Hatefi and Lester in Wisconsin were trying to identify the substances in the electron transport chain acting between NADH and cytochrome *b*. Using lipid extractants they isolated a new quininoid coenzyme which showed redox changes in respiration. They called it coenzyme Q (CoQ). CoQ was later shown to be identical to ubiquinone.

The other type of component to be discovered was the non-heme iron compounds. Iron which was not associated with the porphyrin nucleus had been detected by low temperature electron-spin resonance (Beinert et al., 1959) in mitochondria and seen to undergo changes during respiration. Kearney and Singer showed that highly purified succinate dehydrogenase, a flavoprotein, contained iron. They suggested this non-heme iron might be involved in electron transfer from succinate. Non-heme iron proteins were characterized in bacteria and plants in the early 1960s and found to transfer electrons between flavoproteins and the cytochrome chain and between cytochromes *b* and c_1.

One further advance made by the Wisconsin group was the isolation by the early 1960s of electron transferring complexes. This they achieved by extracting beef heart mitochondria using detergents, sonication, and/or organic solvents. They then reconstructed electron transferring activity between successive complexes. The complexes were mixed in equimolar proportions. Electron transfer from NADH or succinate to oxygen could be reestablished, provided CoQ and cytochrome *c*, both of which are soluble, were added back. Individual components within the particles were then identified.

Work still continues on the way in which electron transfer occurs physically between the different prosthetic groups, the activation of oxygen as acceptor for the final stage in the oxidation, and the association between electron transfer and the generation of ATP (see below).

OXIDATIVE PHOSPHORYLATION

The Concept of High Energy Phosphate

Following the discovery of creatine phosphate and ATP, Meyerhof and Lohmann rather quickly suggested (1931) that energy released from glycolysis was linked to the synthesis of ATP. The phosphate groups in the hexose esters comprised "the unique carriers of the chemical coupling process." The looseness of this phrase was criticized by Parnas (1934) who realized definite partial reactions were likely to be involved, not glycolysis as a whole. By 1941 Lipmann, who had worked with Meyerhof between 1927 and 1932, suggested in his seminal review (*Advances in Enzymology*, Vol.1) that the purpose of catabolism was to utilize energy released by the oxidation of C-H and C-C bonds to generate ATP.

Lipmann distinguished two classes of phosphate derivatives, one with phosphate linked to alcohols, having ΔG of -2 to -3 kcals/mole and a second group with P-O-P as in pyrophosphate, N-P as in creatine phosphate, and O=C-O-P as in 1,3 diphosphoglyceric acid. From the data then available Lipmann calculated the ß-γ phosphate bond in ATP might have a free energy of hydrolysis of about 11 kcals/mole and for PEP, 15.9 kcals/mole. In his article Lipmann introduced the phrase "high-energy phosphate" and its representation as ~P.

The unification and direction which the ~P concept gave to metabolic studies led to its rapid and enthusiastic adoption by the new generation of biochemists starting work after the end of World War II. Kalckar suggested the contribution resonance stabilization might make to the distinction between the two classes of phosphate compounds. T.L. Hill and Morales (1951) identified ionization effects and electrostatic repulsion as further possible contributants to the net differences in free energy between products and reactants.

Some physical chemists vehemently attacked the idea of energy-rich phosphate bonds, implying that the term was frequently being used in its literal sense, irreconcilably with physicochemical principles, and suggestive of "fictitious kinds of bond energy" (Gillespie et al., 1953). While the authors did not quite impute there had been a return to vitalism, the argument stimulated biochemists to a critical evaluation of experimental data, particularly reconsideration of intracellular H^+ concentration, levels of free (unbound) intermediates, Mg^{2+} concentration and possible intracellular compartmentation. Valid equilibrium constants were therefore determined from which free energy changes could be calculated. Krebs and Kornberg in their review, "Energy Transformations in Living Matter" (1957), referred to "substrate level phosphorylation" and the evident involvement of ATP in chemical (biosynthetic) and physical (movement, active transport, bioluminescence) work in cells. [It is now thought that the hydrolysis of ~P is not the source of the energy release. Rather, changes in binding energy, solvation energy and entropy play key roles in energy transduction (de Meis, 1993)].

The complete acceptance of the scheme required that the substantial energy liberated in the oxidation of pyruvate through the tricarboxylic acid cycle ultimately be linked to the formation of pyrophosphate bonds. If this hypothesis was true it was apparent that oxidative phosphorylation must be very different from the phosphorylation occurring in glycolysis. In the breakdown of glucose to pyruvate the substrates themselves were phosphorylated. The further oxidation of pyruvate however involved no obvious phosphorylated intermediates and no apparent dependence on inorganic phosphate (P_i). By the late 1930s, however, the promotion of oxygen uptake by the addition of P_i had been reported by a number of workers.

One of the earliest of these was Engelhardt (1930-1931), initially using mammalian erythrocytes. If glycolysis was blocked by fluoride, reduction of methylene blue still occurred, with esterification of phosphate. With pigeon erythrocytes exposed to alternating periods under oxygen or nitrogen, ATP was alternatively synthesized or hydrolyzed when the conditions became anaerobic. Kalckar (1937) and Belitzer and Tsibakowa (1939) all noted phosphate was esterified when succinate and other dicarboxylic acids were oxidized. Ochoa, then working with Peters and Banga in Oxford, observed ATP formation by brain brei when pyruvate was oxidized, and Cori and his colleagues (1940) concluded "α-keto-acid oxidation in general furnishes energy-rich phosphate bonds."

Both Kalckar and the Russians measured the ratio of atoms of phosphorus yielding ATP to the atoms of oxygen utilized (P/O ratio) and found values significantly greater than 1, the figure expected if a phosphate group had been introduced into a cycle intermediate, as in the glycolytic pathway.

Interpretation of these early experiments with crude tissue preparations was greatly complicated by the presence of very active phosphatases (ATPases) which rapidly hydrolyzed any ATP which might have been formed. Ochoa suggested that the amount of P_i apparently esterified should be corrected for the measurable rate of ATP hydrolysis by the preparation. This gave P/O ratios approaching 3. Belitzer and Tsibakowa and Ochoa realized that "phosphorylation must occur not only when the substrate is dehydrogenated ... but also during the further passage of H (or electrons) along the respiratory chain to O_2." These insights were disregarded, mainly because not all the components of the respiratory chain were yet (ca. 1942-1943) identified and because phosphorylation and oxygen uptake by crude preparations were extremely unstable and irreproducible making quantitative analysis difficult. Theoretical arguments were also advanced (Ogston and Smithies, 1948) against the feasibility of P/O ratios exceeding 2—reasoning that had to be modified when better values for the free energy changes at the different stages in electron transfer became available.

Work recommenced in the late 1940s with an extremely important series of experiments by Lehninger. Initially he and Friedkin used rat liver particles incubated with NADH and $^{32}P_i$ and detected ^{32}P incorporation into ATP if the preparations were aerobic. No incorporation was found if NAD^+ was used as substrate rather than NADH or in the absence of oxygen. That reoxidation of NADH was the direct origin of the energy for phosphorylation was confirmed by Kennedy and Lehninger (1949-1951) using mitochondria whose routine preparation had just become available (see chapter 9). Intact mitochondria are impermeable to NADH but if the mitochondria were made leaky by treatment with hypotonic medium, NADH entered the organelles and was oxidized giving P/O ratios up to 1.89.

This confirmation that phosphorylation was associated with the passage of electrons along the transport chain was immediately appreciated. Work started to identify precisely how high-energy phosphate bonds could be generated and at what steps in the chain. It was soon realized that experimental conditions were very important. Mitochondria could easily be damaged by too vigorous homogenization during their isolation. Exclusion of Ca^{2+} diminished ATPase activation and so gave higher P/O ratios. Trapping the ATP formed as a more stable derivative; e.g. by using glucose and hexokinase to give G-6-P, also increased the P/O value.

With these improved techniques β-hydroxybutyrate, which penetrates mitochondria easily and is oxidized to acetoacetate using NAD^+ as H acceptor, gave a P/O ratio of 3, the value equivalent to that from the reoxidation of NADH found by Lehninger. Succinate, which bypassed the NAD^+/NADH step, gave a ratio of 2. When cytochrome $c{\cdot}Fe^{2+}$ was used, with ascorbate as acceptor, a P/O value tending to 1 was obtained (Judah, 1951). 2-Oxoglutarate oxidation gave a value approaching 4. Since the substrate is oxidatively decarboxylated to yield NADH, which gives a P/O ratio of 3, an additional phosphorylation step must be present—the one substrate level phosphorylation in the tricarboxylic acid cycle (Hunter, 1949). Inhibitors of the cytochrome chain did not affect this phosphorylation.

A different and simpler approach to the measurement of P/O ratios came from the introduction of an oxygen electrode suitable for biochemical studies. Chance and Williams (1955) established conditions under which mitochondrial respiration, in the presence of excess substrate, was totally dependent on the amount of ADP available, i.e., the mitochondria were exhibiting respiratory control. From the change in potential when a known amount of ADP was admitted into the electrode vessel, the oxygen uptake and thus the P/O ratio could be determined, completely confirming the earlier results.

By the mid-1950s, therefore, it had become clear that oxidation in the tricarboxylic acid cycle yielded ATP. The steps had also been identified in the electron transport chain where this apparently took place. Most biochemists expected oxidative phosphorylation would occur analogously to substrate level phosphorylation, a view that was tenaciously and acrimoniously defended. Most hypotheses entailed the formation of some high-energy intermediate X ~Y which, in the presence of ADP and P_i would release X and Y and yield ATP. A formulation of the chemical coupling hypothesis was introduced by Slater in 1953,

$$AH_2 + B + C \rightleftharpoons A \sim C + BH_2 \text{ (i)}$$

$$A{\sim}C + ADP + Pi \rightleftharpoons A + C + ATP \text{ (ii)}$$

and was representative of ideas then under consideration. There was considerable discussion regarding the oxidation state of the postulated high-energy intermediate, which was alternatively regarded as being in the reduced state: BH_2 ~C. Claims were in fact made between the late 1950s and early 1960s for the existence of such derivatives (e.g. NADH ~P) but no confirmation was forthcoming. One of the most persistent, though not logically rigorous, arguments against the high-energy intermediate theory was the continuing inability to detect such intermediates.

Sophisticated isotope experiments were also performed using $H_2{}^{18}O$ (Mildred Cohn) and ^{32}P, and various exchange reactions identified between ATP, ADP, and P_i. Analysis of the mode of action of two inhibitors was also relevant. Dinitrophenol (DNP) uncoupled the association between oxidation and ATP generation (Lardy and Elvejhem, 1945; Loomis and Lipmann, 1948). Oligomycin inhibited reaction (ii) above, blocking the terminal phosphorylation to give ATP, but not apparently the formation of A ~C.

Between 1955 and 1960 various sub-mitochondrial preparations were developed to give vesicles comprising only sealed inner mitochondrial membranes. Cooper and Lehninger used digitonin extraction; Lardy and Kielley & Bronk prepared sub-mitochondrial particles by sonication. At this time, too, Racker and his colleagues isolated F_0/F1 particles from mitochondria and showed that a separated F1 particle behaved as an ATPase. The F_0 portion had no enzymic properties but conferred oligomycin sensitivity on the F1 ATPase. The orientation of these sub-mitochondrial vesicles (inside-out or vice-versa) was shown by the position in electron micrographs of the dense (F1) particles which in normal intact mitochondria project into the matrix and so define the surface of the inner mitochondrial membrane.

The spatial separation between the components of the electron transport chain and the site of ATP synthesis was incompatible with simple interpretations of the chemical coupling hypothesis. In 1964, Paul Boyer suggested that conformational changes in components in the electron transport system consequent to electron transfer might be coupled to ATP formation, the conformational coupling hypothesis. No evidence for direct association has been forthcoming but conformational changes in the subunits of the F1 particle are now included in the current mechanism for oxidative phosphorylation.

By 1960 it seemed likely that vesicles (sealed membrane preparations) were essential for ATP formation to occur. R.J.P. Williams (1959-1961; see 1993) postulated that complex assemblies of catalysts, as in the cytochrome chain, would allow spatial separation of reaction products within the mitochondrial membrane. The initial site of reaction with oxygen, cytochrome oxidase, could be "dislocated" from the site of H^+ generation so that a proton produced at high local concentration in the region of the ATPase could "bring about condensation polymerization such as polyphosphate formation." Mitchell on the other hand, proposed (1961) that ATP was generated consequential to the establishment of a proton/electrochemical gradient across a membrance, an idea unfamiliar to biochemists trained to analyze the molecular participants in scalar reactions. This new concept—the chemiosmotic theory—was strongly resisted by more conservative workers in the field (see Gilbert and Mulkay, 1984) but is now universally accepted. Mitchell received a Nobel prize for his work in 1978.

In his review of the evolution of the chemiosmotic theory between 1961 and 1966, Mitchell (1967) described how his ideas about vectorial translocation of protons across proton-impermeable membranes had been influenced by parallel studies in other systems where ATP was used to move ions against a concentration gradient. Skou in 1957 had identified the Na^+, K^+, and Mg^{2+}-dependent ATPase as the "sodium pump," an entity postulated from isotope and kinetic studies in erythrocytes and central to electrophysiological analyses of nerve membranes. Skou showed ATPase to be asymmetrically situated in the plasma membrane with its cytosolic surface interacting with ATP in a Na^+ -dependent reaction. Potassium ion and ouabain sensitivity were functions of the extracellular face of the enzyme. In 1961, Crane had established that Na^+-dependent glucose uptake in active transport was located at the mucosal, brush-border surface of the intestinal muscoal epithelial cell, spatially separated from energy-dependent Na^+ extrusion (Na pump) in the serosal surface. In mitochondria, Chappell had described anion translocators for dicarboxylic acids and between 1964-1965 Klingenberg and his associates identified the atractyloside-sensitive ATP/ADP transporter in the inner mitochondrial membrane. Significantly Klingenberg's group had also reported conditions in mitochondria *in vitro* when ATP could be shown to drive the electron transport chain "backwards" to generate NADH.

In Mitchell's earliest physicochemical formulations of the chemiosmotic theory, any involvement of the membrane across which the proton gradient was established received little attention. Between 1961 and 1966, Mitchell and Moyle established that at least $2H^+$ were transferred outwards across the membrane for each molecule of ATP generated, although the stoichiometry of H^+ transfer remained controversial. They also suggested the uncoupling of oxidative phosphorylation produced by DNP was due to the phenolate anion acting as a proton trap and so dissipating the proton gradient. The first experiment to demonstrate that a proton gradient across a biological membrane actually did result in ATP generation was that of Jagendorf in 1966 with chloroplasts. Biochemists studying photosynthetic phosphorylation were among the first converts to the chemiosmotic theory.

To explain how H^+ transfer occurred across the membrane Mitchell suggested the protons were translocated by redox loops with different reducing equivalents in their two arms. The first loop would be associated with flavoprotein/non-heme iron interaction and the second, more controversially, with CoQ. Redox loops required an ordered arrangement of the components of the electron transport system across the inner mitochondrial membrane, which was substantiated from immunochemical studies with submitochondrial particles. Cytochrome *c,* for example, was located at the intermembranal face of the inner membrane and cytochrome oxidase was transmembranal. The alternative to redox loops, proton pumping, is now known to be a property of cytochrome oxidase.

Probably the most elegant evidence for the coupling of vectorial H^+ translocation to ATP synthesis came from later experiments of Stoeckenius and Racker who constructed lipid vesicles containing bacterial rhodopsin, which is a proton pump, and beef heart F_0/F1 particles. Light then caused the vesicles to make ATP. Details of the structure of the F_0 unit, through which the effects of the proton gradient are communicated to the synthesizing system, and the way in which the adenine nucleotides and P_i interact with the subunits of the F1 particle are still emerging. Present views indicating that conformational/entropic changes cause ATP formation from bound ADP and P_i, followed by an energy-dependent release of the ATP, are an almost complete reversal of the original chemical coupling hypothesis.

The 1930s were extremely significant in the development of biochemistry. The urea (see Chapter 6) and tricarboxylic acid cycles were described by H.A.Krebs and the terminal pathway of electron transfer outlined by Keilin. Sufficient information was accumulated about the properties of phosphate esters and the formation of ATP to enable Lipmann by 1941 to recognize the metabolic importance of "high-energy phosphate bonds" and to ensure that after the existence of oxidative phosphorylation had been confirmed by Lehninger in the late 1940s, investigations into its mechanism were immediately initiated. Both in the 1920s with the contrasting views of Warburg and Wieland on the nature of oxido-reduction reactions, and between 1960 and 1965 with the controversy surrounding the mechanism of oxidative phosphorylation, the studies were accompanied by vigorous, even virulent controversy hardly discernible from the treatment the topics now receive in standard texts.

Analysis of glycolysis was also largely completed. Guidelines could therefore be formulated for the exploration and evaluation of further metabolic pathways, procedures which were to be greatly facilitated by the introduction of isotopic methods (Chapter 8) which came into general use in the 1950s.

REFERENCES*

Chance, B.& Williams, G.R. (1956.) The Respiratory chain and oxidative phosphorylation. Adv. Enzymol. 17,65-134.

Florkin, M. (1975). A History of Biochemistry, Part III. "Comprehensive Biochemistry (Florkin, M. & Stotz, E.H. Eds.) Vol. 31. Elsevier, Amsterdam.

Gilbert, G.N. & Mulkay, M. (1984). Opening Pandora's Box. Cambridge University Press.

Gillespie, R.J., Maw, G.A., & Vernon, C.A. (1953). The concept of phosphate bond energy. Nature, London, 171, 1147-1149.

Hatefi, Y. (1963.) Coenzyme Q. Adv.Enzymol. 25,275-328.

Holmes, F.L. (1992). Between Biology and Medicine: The Emergence of Intermediary Metabolism. Berkeley.

Holmes, F.L. (1993). Hans Krebs: Architect of Intermediary Metabolism 1933-1937, Oxford University Press, New York and Oxford.

Keilin, D. (1925). On cytochrome, a respiratory pigment common to animals, yeasts and higher Plants. Proc. Roy. Soc. B 98, 312-329.

Krebs, H.A. (1943). The intermediary stages in the biological oxidation of carbohydrate. Advances in Enzymol. 3, 191-252.

Krebs, H.A. (1981). Reflections and Reminiscences. Oxford University Press.

Krebs, H.A. & Johnson, W.A. (1937). The role of citric acid in intermediate metabolism in animal tissues. Enzymologia 4, 148-156.

Krebs, H.A. & Kornberg, H.L. (1957). Energy Transformations in Living Matter. Springer-Verlag, Berlin.

Lipmann, F.(1941). The metabolism, generation and utilization of phosphate bond energy. Adv. Enzymol. 1,99-162.

Lipmann, F. (1953). Development of the acetylation problem: a personal account. In: "Nobel Lectures. Physiology or Medicine, 1942-1962, pp. 413-438. Elsevier (1964), Amsterdam.

Mitchell, P. (1966). Chemiosmotic Coupling in Oxidative and Photosynthetic Phosphorylation. Glynn Research, Bodmin, Cornwall, U.K.

Mitchell, P. (1967). Active transport and ion accumulation. In: Comprehensive Biochemistry (Florkin, M.& Stotz, E.H. Eds.), Vol. 22, pp. 167-197.

Nicholls, P. (1963). Cytochromes: A survey. In: The Enzymes 2nd ed. (Boyer, P.D., Lardy, H., & Myrback, K. Eds), pp. 3-40. Academic Press, New York.

Ogston, A.G. (1948). Interpretation of experiments on metabolic processes using isotopic tracer elements. Nature, London, 162, 963.

Peters, R.A. (1963). Biochemical Lesions and Lethal Synthesis. Pergamon Press, Oxford.

Sayers, D.L. (1930). Strong Poison. Gollancz, London.

Umbreit, W.W., Burris, R.H., & Stauffer, J.F. (1945). Manometric Techniques and Related Methods for the Study of Tissue Metabolism. Burgess Publishing, Minneapolis.

Warburg, O. (1930). The enzyme problem and biological oxidations. Bull. Johns Hopkins Hosp. 46, 341-358.

Williams, R.J.P. (1993). The history of Proton-driven ATP formation. Bioscience Reports 13, 191-212.

Wood, N.A.P. (1978). An inexpensive, simple and robust O_2 electrode. J.Physiol. (London) 275, 3P.

*The tricarboxylic citric acid cycle is shown in Appendix 2

FRIEDRICH WOHLER

THEODOR SCHWANN

LOUIS PASTEUR

EDOUARD BUCHNER

CHRISTIAAN EIJKMAN

FREDERICK GOWLAND HOPKINS

FREDERICK BANTING

ARCHIBALD GARROD

ARTHUR HARDEN

OTTO MEYERHOF

KARL LOHMANN

LUIS LELOIR

CARL CORI

GERTI CORI

TORSTEN THUNBERG

ALBERT SZENT-GYORGI

HANS KREBS

OTTO WARBURG

CHARLES MACMUNN

DAVID KEILIN

FRITZ LIPMANN

FEODOR LYNEN

VLADIMIR ENGELHARDT

PETER MITCHELL

GEORGE HEVESY

RUDOLPH SCHOENHEIMER

DAVID RITTENBERG

MICHAIL TSVETT

JAMES SUMNER

THE SVEDBERG

STANFORD MOORE

WILLIAM STEIN

Chapter 6

AMINO ACID CATABOLISM IN ANIMALS

AMINO ACID CATABOLISM: THE ROLE OF THE LIVER

By the middle of the nineteenth century it was generally accepted that nitrogen was excreted by mammals as urea, and by 1905 Folin had shown that the amount of urea voided through the kidneys as urine was proportional to the level of protein in the diet. Normally 20-30 g urea are excreted by man per day; protein intake also affects urine volume, which is usually 1.2-1.5 L per day.

That the liver is the main site of nitrogen catabolism in higher vertebrates had been inferred from patients with liver disorders and was confirmed once levels of urea in blood and urine could be determined accurately. In 1924 Bollman, Mann and Magrath demonstrated that hepatectomized dogs were unable to make urea. Another important observation was that catabolism of amino acids was oxidative. People with alkaptonuria (Garrod, Chapter 3) who were unable to metabolize tyrosine correctly, excreted homogentisic acid in their urines, a metabolite Neubauer (1909) realized had been derived from tyrosine following oxidative deamination. He and Knoop independently suggested deamination might occur with the formation of α-ketonic acids.

Some experiments had already indicated the fate of the C skeletons of the amino acids. Kossel (1898) was the first to suggest protein could be catabolized to glucose. This was proved by Stiles and Lusk

(1903) who injected dogs with phloridzin which selectively blocked glucose reabsorption in the renal tubule. When an amino acid mixture was fed to the dogs its nitrogen content appeared in the urine as urea, with the C skeletons as glucose. Experiments with perfused livers also confirmed this organ as the site for urea production in mammals. If the livers were taken from starved animals, so that the organ was initially glycogen free, amino acids in the perfusate were deaminated and glycogen deposited. A few amino acids promoted acetoacetate formation leading to the distinction between glycogenic and ketogenic amino acids.

Arginine was isolated by Schulze in the 1890s. That it might be the immediate precursor of urea was suggested from observations by Kossel and Dakin early this century when they saw how rapidly isolated liver autolyzed, yielding urea. They identified the enzyme responsible, as arginase. The second product of arginase activity, ornithine, was identified by Schulze. The presence of arginase in liver was recognized by Clementi (1914) as the critical enzyme distinguishing ureotelic from uricotelic vertebrates. Mammals, aquatic amphibia, and elasmobranch (cartilaginous) fishes use urea as the main product of nitrogen metabolism. Birds and reptiles, where development occurs in a strictly limited water supply, have cleidoic eggs and use uric acid as the excretory product (Joseph and Dorothy Needham, 1932). They do not have arginase in their livers. By 1930, McCance and his colleagues at the Nutrition Unit in Cambridge had calculated that in humans all the urea excreted was derived via the breakdown of liver arginine.

The experiments described above indicated amino acids were oxidatively deaminated in liver and their α-amino groups converted to urea. A start on investigations of the mechanism of urea biosynthesis was made by Schultzen and Nenki (1869) who concluded that amino acids gave rise to cyanate which might combine with ammonia from proteins to produce urea. Von Knieren (1873) demonstrated that when he drank an ammonium chloride solution, or gave it to a dog, there was an increase in the formation of urea, without any rise in urinary ammonia. His results were consistent with the cyanate theory but did not eliminate the possibility that urea arose from ammonium carbonate which could be dehydrated to urea:

$$2NH_3 + CO_2 + H_2O \rightarrow (NH_4)_2CO_3 \rightarrow CO\begin{matrix} NH_2 \\ ONH_4 \end{matrix} \rightarrow CO(NH_2)_2$$

ammonium carbonate ammonium carbamate

Feeding experiments showed that this reaction did not occur biologically nor, in contrast to the Wohler experiment, was ammonium cyanate converted to urea in the liver. Indeed cyanic acid, (HCNO), was very toxic.

Waldemar von Schroder (1882) perfused dog liver with ammonium chloride and like von Knieren and Salkowski with intact animals, obtained urea. Numerous investigators employed perfusion in the next decades; the addition of amino acids to the perfusing blood gave increased amounts of urea.

EXPERIMENTAL PROCEDURES

Estimation of Urea

The hypobromite method for the quantitative estimation of the amount of urea in urine was developed about 1870. Alkaline hypobromite released nitrogen from urea, which was measured volumetrically:

$$3\ NaOBr + CO(NH_2)_2 + 2\ NaOH \rightarrow N_2 + 3NaBr + Na_2CO_3 + 3H_2O$$

The enzyme urease was discovered in soybeans by Takeuchi in 1909; it catalyzed the conversion of urea to ammonium carbonate. Jack beans were another excellent source of the enzyme. Jack bean powder could be stored for considerable periods and very active, soluble, urease extracted. After the action of urease, the ammonia could be estimated colorimetrically by Nesslerisation or titrimetrically. The Conway diffusion apparatus was specially developed for the estimation of urea titrimetrically and remained in use into the 1950s.

In 1927, Van Slyke devised a micromethod using urease. Instead of determining ammonia released by alkali from ammonium carbonate, carbon dioxide was liberated after acidification. This was successfully adapted by Krebs to Warburg manometry, the carbon dioxide being released at pH 5 by acetate buffer. It therefore became possible to carry out multiple kinetic experiments rapidly under identical conditions.

Tissue Preparations

Liver perfusion was devised by Bernard (1855) for his studies on glucose production. Perfusion was the first procedure which allowed metabolic events to be studied in isolated organized tissue. Alterations in the composition of the perfusate could be followed over time; the surviving organ was available for analysis at the end of the experiment. Weighing the organ offered a simple method for detecting permeability changes and the entry of water. With the development of impermeant solutes like inulin it became possible to discriminate between intracellular and extracellular compartments.

One of the major difficulties in perfusion studies was the maintenance of tissue oxygenation and the avoidance of "outflow block" which led to swelling, congestion, and serious restrictions in the flow through the liver (Kestens, 1964). These difficulties were hard to overcome with early procedures which used oxygenated salt solutions as perfusates. By the early 1940s Whipple and Madden using oxygenated homologous blood reported levels of plasma protein formation in perfused dog livers comparable to those occurring in intact animals. In the 1940s perfusion studies were adapted to the organs of small animals. With large animals it was necessary to perfuse the liver through both the portal vein and the hepatic artery; with rats only the simpler portal inflow was required.

In 1951, L.L. Miller developed an artificial "lung" where a thin film of heparinized blood came into contact with humidified air or oxygen. With this apparatus Miller and his associates confirmed the central role of the liver in the synthesis of plasma proteins. In the mid-1960s Krebs, Hems, and their colleagues commenced very successful studies on gluconeogenesis in perfused livers, monitoring the performance of the organ by the high rates of urea formation from ornithine and ammonium chloride. Technological innovations, such as the availability of fine plastic tubing which was used to cannulate the vessels, also greatly facilitated the procedure (R. Hems, personal communication). With the improved perfusion technique it was possible to study the behavior of small populations of cells within the organ; for example, in liver the bile duct could be canulated and the behavior of the bile duct epithelium observed.

Perfusion experiments before oxygenation problems were solved often gave incomplete or misleading information. Embden's work on fatty acid oxidation in liver (see Chapter 7) suggested acetoacetate was a normal metabolite, although in the intact animal free acetoacetate is only detected when glucose oxidation is impaired.

THE UREA CYCLE

The seminal contribution of Krebs and Henseleit was to show that ornithine is catalytic in the formation of urea from ammonia. Krebs had moved to Freiburg in the spring of 1931, and had set up his own laboratory in Thannhauser's department. Kurt Henseleit, a medical student, was his first research student. Henseleit carried out much of the work on the synthesis of urea under Krebs' detailed guidance. Initially the experiments were directed to the identification of amino-containing compounds which might be precursors of urea and /or might accelerate the formation of urea from ammonium chloride. Liver slices were used because they allowed many tests to be performed with the same organ. The formation of urea was measured in Warburg manometers and could be compared under various conditions.

In October 1931, Henseleit noted the enhancement in urea production caused by the addition to the system of arginine. The effect of ornithine was first observed in November that year (Holmes, 1991). Krebs' recollections were that "I took ornithine really because it was there." ... "I think it was really... an unprejudiced collections of facts." Control experiments by Henseleit showed ornithine without added ammonium chloride gave no urea—the effect of ornithine was to augment urea formation from ammonium chloride. Ornithine was unique among the large number of compounds tested. Further experiments involved varying the medium and, very significantly, testing the effects of diminishing concentrations of ornithine. Direct determination of ammonia utilization suggested that ammonia was the sole source of nitrogen in urea; little or no nitrogen originated from low concentrations of ornithine itself.

However Henseleit showed there was a correlation between the concentration of ornithine and the magnitude of the effect on urea production. That the absolute amounts of ornithine were so small provided Krebs with arguments for the idea that the ornithine acted catalytically.

The paper by Krebs and Henseleit (1932), "Experiments on the Formation of Urea in Animal Bodies" (*Klinische Wochenschrift* 11,759), contained the phrases…"in the synthesis of urea in the living cell, ornithine acts like a catalyst." "We therefore draw the conclusion …that the primary reaction for the synthesis of urea from ammonia is

$$\text{ornithine} + 2NH_3 + CO_2 = \text{arginine} + 2H_2O$$

$$\text{arginine} + H_2O = \text{ornithine} + \text{urea"}$$

In 1930 Wada isolated citrulline from watermelon. It was an obvious candidate as an intermediate in the urea cycle. When tested by Krebs it was catalytic only at high concentrations. Citrulline is not normally detectable in liver.

The elegance and definitive nature of the paper in *Klinische Wochenschrift* was immediately appreciated. Gowland Hopkins in an anniversary address to the Royal Society (November, 1932) singled out Krebs' research on urea as an outstanding contribution from a young scientist. "The data were obtained by the methods of microanalysis. …The high accuracy to be obtained…by micromethods is now well recognized, but it is becoming clear that technique is so developing that kinetic studies can be made equally accurately on a similar scale. To studies on living systems this offers advantages which cannot be overestimated."

It was not until 1933 after Krebs had decided to leave Freiburg that he formulated urea synthesis in its now common cycle form:

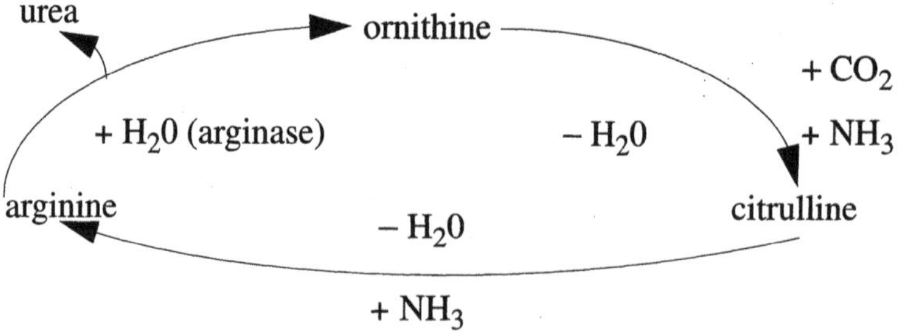

This first appeared in Hildegard Mandershied's M.D. thesis, "On the Formation of Urea in Vertebrates." (Holmes, 1991).

The urea cycle was confirmed by one of the earliest experiments in animals using labeled $^{15}NH_3$ (Schoenheimer and Rittenberg, 1939). The ^{15}N label was found only in urea and in the amidine group of arginine. There was no ^{15}N in the terminal, γ-NH_2 of ornithine. Conversely if α or γ ^{15}N ornithine was used no ^{15}N appeared in urea. That all the carbon was derived from plasma HCO_3^- was confirmed by du Vigneaud (1948) using ^{14}C.

Details of the enzymic interconversions between the intermediates depended principally on P.P. Cohen's and Sarah Ratner's work in the late 1940s and early 1950s. It was greatly helped by the discovery by Cohen and Hayano (1946-1947) that the different stages of the cycle occurred in different parts of the cell. If glutamate and ATP were added to the medium, Cohen and Hayano found that homogenates could be used rather than slices. This allowed them to fractionate the homogenate into the particulate (mitochondrial) and soluble fractions. The particulate fraction catalyzed the formation of citrulline from ornithine and the cell sap contained arginase and the enzymes responsible for the formation of arginine from citrulline.

Cohen and Grisolia then concentrated on the first step in the reaction, obtaining citrulline from ornithine. The reaction appeared to depend on oxygen, a requirement traced to the need for high concentrations of ATP. Physiologically the formation of urea occurs at very low levels of ammonia, which is extremely toxic as it is also lipid soluble and enters cells very easily. Cells are not very effectively buffered against OH^-.

For urea formation *in vitro* at biologically plausible concentrations of ammonia, glutamate had to be present in the medium. This was first ascribed to the need for carbamoyl glutamate to act as ammonia carrier, but later experiments with isotopically labeled $^{15}NH_3$ and $^{14}CO_2$ showed neither the carbonyl nor the amine group of the carbamoyl glutamate was utilized to give urea.

It presently became clear that ammonia, carbon dioxide, and ATP were forming an active intermediate which condensed with ornithine to give citrulline. The clue to its identification came from Lipmann's laboratory in 1955 (Jones, Spector, and Lipmann). Lipmann had been studying a bacterial reaction in which citrulline was metabolized, in the presence of ADP, to give ATP. A new high-energy compound, carbamoyl phosphate, ($H_2N\text{-}CO\text{-}O\text{-}PO_3^{2-}$) was detected as an intermediate. It was very easily synthesized *in vitro* by incubating 1M KH_2PO_4 and 1 M KCNO at 30 °C. After 0.5 h about 50% conversion to carbamoyl phosphate occurred and the relatively stable, acid-labile phosphorus compound could be precipitated as the lithium salt. Liver mitochondria, when incubated with carbamoyl phosphate and ornithine, gave citrulline.

Lipmann and his colleagues confirmed that for carbamoyl phosphate to be synthesized by liver mitochondria, the presence of glutamate was essential, although glutamate itself did not participate in the reaction. Later work showed the glutamate was active as *N*-acetyl glutamate—one of the earliest examples of allosteric control of an enzyme. *N*-acetyl glutamate promotes carbamoyl phosphate synthase by allowing carbamoyl phosphate formation to occur at lower concentrations of ammonia.

Between 1947 and 1949, Ratner and Petrack examined the formation of arginine from citrulline in the post-mitochondrial supernatant. The NH_2 group originated from the α-amino-group of aspartate. Two enzymes were found to be involved, a condensing enzyme which yielded arginino-succinate and succinate lyase which released arginine. Initially the second product was believed to be malate, but when highly purified preparations of the lyase became available, uncontaminated by enzymes of the tricarboxylic acid cycle, fumarate was shown to be the correct end-product.

An anomaly associated with citrulline that became evident when detailed kinetic studies were made in the 1950s (R.B. Fisher and J.R. Bronk) was the irreproducibility of its catalytic activity in liver slices on the formation of urea, despite the clear evidence from Ratner and Petrack of its importance in arginine synthesis. Initially the discrepancy in catalytic activity between ornithine and citrulline was ascribed to the possible impermeability of the liver cell plasma membrane to the latter intermediate, a hypothesis which was rapidly disproved experimentally. Only recently has it been shown that ornithine transcarbamylase is clearly associated with the ornithine/

citrulline transporter in the inner mitochondrial membrane. During synthesis of ^{3}H-citrulline from ^{3}H-ornithine, radioactivity from the substrate does not mix with unlabeled ornithine in the mitochondrial matrix but is thought to be directly channeled from its transporter to the carbamlytransferase (see Meijer et al., 1990). The citrulline thus synthesized is assumed then to be translocated to the active site of arginino-succinate synthetase (Cheung et al., 1989), so precluding equilibrium with exogenous citrulline—an extension to the need for proper substrate accessibility to its enzyme which Krebs had identified in his review of conditions necessary to establish the existence of a pathway (see Chapter 5).

AMINO ACID OXIDATION AND THE RELEASE OF AMMONIA

Besides investigating the reactions by which ammonia was converted to urea, Krebs also turned his attention to the origins of the ammonia. An oxidase was discovered (1932-1933) which catalyzed oxidative deamination:

$$R{\cdot}CHNH_2{\cdot}CO_2H + O_2 \rightarrow R{\cdot}C{=}NH.CO_2H + H_2O_2$$

$$R{\cdot}C{=}NH{\cdot}CO_2H + H_2O \rightarrow R{\cdot}CO{\cdot}CO_2H + NH_3$$

Activity was particularly evident in kidney slices.

Many of the amino acids originally tested by Krebs were racemic mixtures. When naturally occurring L-amino acids became available the oxidase was found to be sterically restricted to the unnatural, D series. [D-serine occurs in worms free and as D-phosphoryl lombricine (Ennor, 1959)]. It could not therefore be the enzyme used in the liver to release NH_3 in amino acid metabolism. D-amino acid oxidase was shown by Warburg and Christian (1938) to be a flavoprotein with FAD as its prosthetic group. A few years later Green found an L-amino acid oxidase in liver. It was however limited in its specificity for amino acid substrates and not very active—characteristics which again precluded its central role in deamination.

A further amino acid oxidizing enzyme, utilizing glutamate as substrate, which was not a flavoprotein but used pyridine nucleotide coenzymes as H-acceptor, was found by von Euler (1937) in the particulate fraction of cells. Glutamate dehydrogenase was also present in plants, a wide range of animal tissues, and in microorganisms, and was extremely active. In animals it used NAD^+ as H-acceptor and is now known to make up about 5% of mitochondrial protein. The enzyme catalyzed reversible oxidative deamination of L-glutamic acid to yield 2-oxoglutarate, ammonia, and NADH which was reoxidized by the mitochondrial electron transport chain. Since 2-oxoglutarate was an intermediate in the tricarboxylic acid cycle, the link between amino acid and carbohydrate metabolism was established.

TRANSAMINATION

The discovery of L-glutamate dehydrogenase explained how glutamic acid could be oxidized to release ammonia. It did not immediately account for the deamination of other amino acids. A clue to their catabolism was provided by Dorothy Needham (1930) who noted that in pigeon breast muscle, glutamate and aspartate disappeared without a concomitant decrease in amino-N. Amounts of urea and ammonia did not increase, but succinate levels rose. She concluded percipiently: "Possibly a combination of the amino group with some reactive carbohydrate residue takes place ... [and] the amino group is retained in the form of a new amino acid." Sundry other observations were made which retrospectively could be interpreted as evidence for such an amino-group transfer. Szent-Gyorgi, for example, again using pigeon breast muscle, saw that the addition of glutamate markedly promoted oxaloacetate disappearance.

In 1937 Braunstein and Kritzman reported the presence in minced pigeon breast muscle of a new enzyme system, a transaminase catalyzing the transfer of an α–amino group from glutamic acid to an accepting α-keto acid, pyruvate, yielding 2-oxoglutarate and alanine. Braunstein immediately recognized how the existence of this reaction could account for deamination, if transaminases existed which could be coupled to L-glutamate dehydrogenase.

Demonstrating the existence of multiple transaminases in the late 1930s and early 1940s was very difficult. Chromatographic separation of amino acids was not available until the mid-1940s and paper chromatography was not in common use until 1948-1950 when Hird and Rowsell finally showed the existence of the wide range of transaminases and the universality of the transamination process.

Work in the preceding 10 years had however been simplified by the development by P.P.Cohen of a new, specific micromethod for the estimation of glutamate, following a suggestion of Krebs. Glutamic acid specifically reacts with chloramine—T, an active form of hypochlorite, to give succinonitrile and carbon dioxide which could be measured manometrically.

It was then possible to confirm the existence of two transaminating systems, the original one utilizing pyruvate as amino acceptor, and a second which used oxaloacetate. Both enzymes were purified and found to be very specific for their substrates. The reactions catalyzed were freely reversible.

The biological importance of transamination was confirmed using ^{15}N-labeling experiments (Tannenbaum and Shemin, 1950). ^{15}N-leucine incubated with pig heart muscle gave highly labelled ^{15}N-glutamate, evidence that leucine could be transaminated. Isotope experiments were then extended to the whole range of amino acids.

PYRIDOXAL PHOSPHATE (VITAMIN B_6) AS COENZYME FOR TRANSAMINATION

The association between vitamin B_6 deficiency and transamination emerged from 1945 when Schlenk and Fisher noted that pyridoxine-deficient rats had a diminished capacity for transamination. In the same year Gunsalus and his colleagues found transamination in *Streptococcus faecalis* depended on pydridoxal phosphate. The properties of the heat-stable component in purified glutamic-oxaloacetate transaminase were similar to those of pydridoxal phosphate. Later pyridoxal phosphate was established as an essential coenzyme in many amino acid transformations.

Back in 1937, Braunstein and Kritzman and, independently, Herbst, had proposed transamination might proceed via a Schiff's base formation. The essential lability of the H atom on the α-C atom was shown with deuterium labelling (1942). α-^2H alanine released ^2H into the medium during transamination. The label did not appear in glutamate, the end-product.

Identification of pyridoxal phosphate as coenzyme suggested the aldehyde group on pyridoxine might form an intermediate Schiff's base with the donor amino acid. Pyridoxamine phosphate thus formed would in turn donate its NH_2 group to the accepting α-ketonic acid, a scheme proposed by Schlenk and Fisher. ^{15}N-labeling experiments and, later, the detection of the Schiff's base by its absorption in UV, confirmed the overall mechanism. Free pyridoxamine phosphate however does not participate in the reaction as originally proposed. Pyridoxal phosphate is invariably the coenzyme form of pyridoxine.

A few years later, in 1953, the versatility of pyridoxal phosphate was illustrated by Snell and his collaborators who found many of the enzyme reactions in which pyridoxal phosphate is a coenzyme could be catalyzed non-enzymically if the substrates were gently heated with pyridoxal phosphate (or free pydridoxal) in the presence of di- or tri-valent metal ions, including Cu^{2+}, Fe^{3+}, and Al^{3+}. Most transaminases however are not metal proteins and a rather different complex is formed in the presence of the apoprotein.

REFERENCES*

Baldwin, E. (1949.) An Introduction to Comparative Biochemistry. Cambridge University Press.

Cheung, C.W., Cohen, N.G., & Raijman, L. (1989). Channeling of urea cycle intermediates *in situ* in permeabilized hepatocytes. J. Biol. Chem. 264, 4038-4044.

Cohen, P.P. (1954). Nitrogen metabolism of amino acids. In: Chemical Pathways in Metabolism (Greenberg, D.M., Ed.), Vol. 2, pp. 1-46. Academic Press, New York.

Fisher, R.B. (1954). Protein Metabolism. Methuen, London.

Holmes, F.L. (1991) Hans Krebs: The Formation of a Scientific Life 1900-1933. Vol.I. Oxford University Press, New York and Oxford.

Jones, M.E., Spector, L., & Lipmann, F. (1955). Carbamoyl Phosphate. In: Proceedings of the 3rd International Congress of Biochemistry. (1956) (Liebecq, C., Ed.) Vaillant-Carmanne, Liege.

Kestens, P.J. (1964). La Perfusion du Foie Isole. Editions Arscia S.A. Bruxelles.
Meyer, A.J. Larners, W.H., & Chamulead, R.A.F.M. (1990). Nitrogen metabolism and ornithine cycle fixation. Physiol. Rev. 70, 701-748.

*The urea cycle is shown in Appendix 2

Chapter 7

THE UTILIZATION OF FATTY ACIDS

CLASSICAL FATTY ACID OXIDATION

Fatty acid oxidation was the first biochemical pathway to be understood in outline. Wohler in 1824 showed fatty acids could be totally oxidized to carbon dioxide and water. Respiratory quotient determinations (Chapter 3) on animals and humans in the post-absorptive state gave a value of 0.7 as would be predicted if fatty acids were completely oxidized. Gerhardt (1865) had identified acetoacetate in the urines of patients who were diabetic. That the "ketone bodies" in urine—acetoacetate, its reduction product ß-hydroxybutyrate, and acetone—were all derived from the metabolism of fatty acids was suggested by Geelmuyden (1897). It was assumed that these ketone bodies were formed during catabolism but did not accumulate when glucose was being oxidized normally, an interpretation giving rise to the aphorism "Fat burns in the flame of carbohydrate." Another point noted was that in naturally occurring fatty acids even numbers of carbon atoms predominated.

Insight into the process of fatty acid oxidation came from the work of Knoop (1904). He synthesized a series of fatty acids with a phenyl group in the terminal, ω, position. These were then fed to dogs and the urines collected. Phenyl groups were not broken down so that the phenyl "label" remained unchanged when fatty acids were oxidized, and could be detected in the urine. With increasing chain length only two alternative derivatives were detected, hippuric acid derived from benzoic acid after conjugation with glycine, and phenylaceturic acid from phenylacetic acid. Knoop therefore concluded the fatty acids had been catabolized by splitting off two carbon atoms at a time from the carboxyl end of the molecule.

Further studies by Dakin (1904) showed that if phenyl propionic acid, $C_6H_5CH_2CH_2CO_2H$, was fed to cats or dogs, $C_6H_5COCH_2CO_2H$, $C_6H_5CHOHCO_2H$ and $C_6H_5CH{=}CHCO_2H$ could all be detected. The order in which these products were metabolized from the phenyl-labeled propionic acid was not clear, but it was concluded that fatty acid oxidation proceeded by a ß-oxidative attack, two carbon atoms at a time being split off from the carboxyl end of the molecule. Normally the process would continue, to yield finally carbon dioxide and water, but if carbohydrate oxidation was impaired, as in diabetes, the last four carbon atoms of the fatty acid chain gave acetoacetic acid, 1 mole/mole original fatty acid. Model experiments using hydrogen peroxide to oxidize butyric acid *in vitro* showed acetoacetate was formed, again supporting ß-oxidation.

Between 1906 and 1908 the breakdown of fatty acids to acetone was detected by Embden in perfused livers. Only fatty acids with even numbers of carbon atoms produced this effect. The acetone was postulated to have originated from acetoacetate. For the next 30 years the ß-oxidative route of fatty acid oxidation was generally unchallenged. By 1935-1936 however much more accurate determinations of the yields of acetoacetate per mole of fatty acid consumed (Deuel et al., Jowett and Quastel) indicated convincingly that more than one mole of acetoacetate might be obtained from 6C or 8C fatty acids. (Octanoic acid was often used as a model fatty acid as it is the longest fatty acid which is sufficiently soluble in water at pH 7.0 for experimental purposes.) The possibility had therefore to be entertained that 2C fragments could recondense (MacKay et al. 1942).

Although sequential ß-oxidation from the carboxyl end of fatty acids was believed to be the mechanism for their breakdown, other schemes had been proposed, notably by Hurtley in 1915, who suggested multiple alternate oxidation—this idea was not widely accepted because the probable intermediates, polyketonic or polyunsaturated fatty acids, had never been detected. The abnormally high levels of acetoacetate produced by various liver preparations, however, caused multiple alternate oxidation to be reassessed. ^{13}C labeled octanoic acid was therefore used by Weinhouse, Medes, and Floyd (1942) to decide which pathway was being followed. Liver slices were incubated with octanoic acid labeled only in the CO_2H group. The acetoacetate produced was degraded to acetone and carbon dioxide and the isotope content of the two compounds measured. Equal amounts of ^{13}C were found in the carboxyl and carbonyl carbon atoms, a finding apparently incompatible with either the classical ß-oxidation theory of Knoop and Dakin or the multiple alternate oxidation hypothesis.

THE FATTY ACID SPIRAL

Fifty years were to pass before it became possible to define the individual enzyme steps in the ß-oxidative process outlined above. No intermediates in the pathway could be found *in vivo* nor was it possible to detect them in the isolated systems then in use, such as tissue slices. Simpler preparations were needed before the details of the enzymology could be established.

A start to this came in 1939 when Leloir and Munoz obtained a rather unstable particulate, cell-free preparation from liver which, if supplemented with magnesium, AMP, P_i, fumarate, and cytochrome *c*, catalyzed complete fatty acid oxidation rather than producing acetoacetate. When isolated mitochondria became available in the 1940s (Chapter 9), Schneider and Potter, Lehninger and Kennedy, and Green (with cyclophorase, Chapter 9), also reported conditions under which octanoic acid was completely oxidized. The process required ATP and the presence of the tricarboxylic acid cycle intermediates. The use of mitochondria also prevented a major complication which could arise when liver slices were used—the simultaneous presence of both the fatty acid catabolizing and synthesizing systems. Although clinical considerations and the perfusion experiments of Embden had shown the liver to be the main site of fatty acid oxidation, by 1946 Lehninger had also demonstrated that other tissues, especially heart muscle, could utilize long-chain fatty acids very effectively.

In 1942, Lynen found acetate oxidation in starved yeast required the addition of energy which could be provided from the oxidation of ethanol. After his discovery of the need for coenzyme A (CoA) in acetylation reactions (Chapter 5), Lipmann suggested acetyl CoA might be the missing 2C fragment or "active acetate" (Lipmann, Harvey Lectures 1948-1949).

The actual pathway by which fatty acid oxidation occurred was established by Lynen (1952-1953). Its unique and characteristic reaction was the thioclastic attack by coenzyme A on the ß-ketoacyl CoA derivative, splitting off the 2C fragment, acetyl CoA. Free coenzyme A was very difficult to isolate and although it was synthesized in Todd's laboratory in Cambridge in the mid-1950s, much of the early work from Lynen's laboratory utilized *N*-acetyl cysteamine as a not very efficient (ca. 1%) coenzyme A analogue. It carried the essential thiol group of the ß-mercaptoethylamine end of CoA and could be used in most, but not all, of the steps in the spiral.

Lynen had studied chemistry in Munich under Wieland; his skill as a chemist led to the successful synthesis of a number of fatty acyl CoA derivatives which proved to be substrates in the catabolic pathway. Many of these C=O or C=C compounds had characteristic UV absorption spectra so that enzyme reactions utilizing them could be followed spectrophotometrically. This technique was also used to identify and monitor the flavoprotein and pyridine nucleotide-dependent steps. Independent evidence for the pathway was provided by Barker, Stadtman and their colleagues using *Clostridium kluyveri*. Once the outline of the degradation had been proposed the individual steps of the reactions were analyzed very rapidly by Lynen, Green, and Ochoa's groups using in the main acetone-dried powders from mitochondria, which, when extracted with dilute salt solutions, contained all the enzymes of the fatty acid oxidation system.

The mechanism of the initial, cytoplasmic, activation of the fatty acids was established in Lipmann's laboratory:

$$\text{Enzyme + ATP} \leftrightarrow \text{Enz-AMP + PP}$$

$$\text{Enz-AMP + HS-CoA} \leftrightarrow \text{Enz-S-CoA + AMP}$$

$$\text{Enz-S-CoA + acetate} \leftrightarrow \underset{\text{acetyl CoA}}{CH_3\text{CO-S-CoA}} + \text{Enzyme}$$

With highly purified enzyme, free from pyrophosphatase, radioactive exchange between ^{32}PP and ATP could be demonstrated. Later it was shown that at least three different enzymes activated fatty acids, depending on the chain lengths of the substrate.

FATTY ACID SYNTHESIS

The Reversibility of the Fatty Acid Spiral

As early as 1842 Liebig had suggested carbohydrate might be the precursor of fat, a hypothesis conclusively validated by Gilbert and Lawes (1860-1870) at the Agricultural Research Station at Rothamsted. Pigs fed either a high carbohydrate or a high protein diet laid down fat in their carcasses. That fat deposition might be a dynamic process was appreciated after the experiments of Schoenheimer and Rittenberg (1936) with mice fed whole wheat bread and given ^{2}H-enriched water to drink (Chapter 8). The deuterium was incorporated into body fat. Similar experiments by Stetten and Boxer (1944) led those authors to conclude fat synthesis was a more important pathway in the utilization of glucose than its conversion into glycogen. When universally labeled ^{14}C glucose was fed to mice 12.2% of the ^{14}C was present in fatty acids after 24 h and 14.5% after 48 h (Masoro, Chaikoff, and Dauben, 1949).

The importance of acetate in fatty acid synthesis was demonstrated in 1926 by Ida Smedley MacLean and Hoffert. Yeast in a medium with acetate as sole source of carbon grew normally and deposited lipid. In the early 1940s Bloch and Rittenberg showed ^{14}C-acetate was a precursor of fatty acids, the carbon of the methyl group always being found in the terminal, ω, position as well as elsewhere, a conclusion elegantly confirmed by experiments of Popjak with lactating rabbits and goats. When udders were perfused with 1-^{14}C acetate the label was incorporated into carbons 1, 3, and 5 of caproic acid in the milk. Fatty acid synthesis was also examined in microorganisms. If cell-free extracts from *Clostridium kluyveri* were maintained anaerobically, ethanol was converted into butyrate and caproate. In the presence of oxygen, fatty acid breakdown occurred to give acetyl phosphate

(Stadtman and Barker, 1949)—results apparently consistent with the reversibility of the oxidative and synthetic processes.

Reversibility was a very important feature of the fatty acid spiral as originally described. From his early observations Lynen had suggested that fatty acid synthesis occurred by the reactions of the spiral proceeding backwards. Such a proposal seemed inherently probable. The nutrition experiments of the nineteenth century showed that excess carbohydrate was retained in the body as fat. The labeling experiments of Bloch and Rittenberg showed acetate was incorporated into fatty acids *in vivo.* Weinhouse, Medes and Floyd (above) found 2C fragments recondensed to acetoacetate. Further, in the tricarboxylic acid cycle the reactions succinate ↔ fumarate ↔ malate ↔ oxaloacetate were reversible and were formally analogous to the mitochondrial reactions in the fatty acid spiral, RCH_2CH_2CO-CoA ↔ RCH=CHCO-CoA ↔ $RCHOHCH_2CO$-CoA ↔ $RCOCH_2CO$-CoA. Experiments in other laboratories showed that acetyl CoA incubated with ß-ketothiolase gave acetoacetate *in vitro.*

There were also less concrete considerations. In the early 1950s glycogenolysis was still believed to be completely reversible. UTP dependency and the glycogen synthase reactions had not yet been discovered nor had phosphofructokinase been shown to act irreversibly. The mechanism of protein synthesis was still a mystery. Laboratories studying proteolysis had shown that the peptide bond could be resynthesized by peptidases, although under very restricted conditions. Reversibility seemed to be an accepted property of the major metabolic pathways.

Numerous experiments were then performed with mitochondria incubated with acetate, CoA, ATP, etc., in attempts to detect fatty acid synthesis. In 1957, Lynen and his colleagues reported the presence in mitochondria of a system which catalyzed the elongation of caproyl CoA to octanoyl CoA by the addition of an acetate unit. NADH and NADPH had to be present. The existence of this mitochondrial system was confirmed by Wakil et al. in 1961 who showed the 12C acid could be extended to 16C by successive additions of 2C fragments.

In 1955, Wakil however had found a totally different mechanism for fatty acid synthesis which was dependent on the presence of HCO_3^- and was catalyzed by soluble enzymes in the post-mitochondrial supernatant, quite distinct from those in the catabolic route. The initial observation was serendipitous. One weekend normal stocks of phosphate buffer had run out and bicarbonate/carbon dioxide was substituted very successfully. The importance of bicarbonate had already been suggested from experiments of Chaikoff and his associates (1950) who detected fatty acid synthesis from ^{14}C glucose when liver slices were incubated in a bicarbonate-Ringer solution. That synthesis might be catalyzed by a post-mitochondrial supernatant had also been indicated by the observation of Brady and Gurin (1950-1952) that the 10^5-g supernatant from pigeon liver homogenates could synthesize fatty acid if supplemented with a mitochondrial extract. ATP, CoA, and NAD^+ were also required. From 1955, Wakil, Ganguly, Gibson, and their colleagues started to characterize the bicarbonate-dependent route. Kinetic studies then showed the equilibrium of ß-ketothiolase was firmly in favor of the thioclastic split, so precluding physiological reversibility of the spiral on thermodynamic grounds.

Acetyl CoA Carboxylase

Experiments with $^{14}CO_2$ showed very clearly that carbon dioxide was essential for fatty acid synthesis but that the ^{14}C did not appear in the final product. Malonyl CoA was identified in 1958-1959 when two stages in fatty acid synthesis were distinguished, the fixation of carbon dioxide to give malonyl CoA and the buildup of the long-chain fatty acid. Formation of malonyl CoA was thought to be the rate-limiting step (Lynen et al. 1961). By this time mechanisms of carbon dioxide fixation in animal tissues, especially liver, had been quite extensively studied. Avidin inhibited one class of carboxylases, which was then proved to be biotin-dependent (see Chapter 3). Wakil, Gibson, and others found the formation of malonyl CoA to be blocked by avidin and showed biotin to be a component in the carboxylase.

The activating effect of di-and tricarboxylic acids on fatty acid synthesis was first noted by Brady and Gurin with citrate. The stimulation was confirmed when purified acetyl CoA carboxylase became available and was activated by tricarboxylic acids. Vagelos et al. found citrate to be the most effective. They noted that the carboxylase from adipose tissue aggregated in the presence of citrate—an observation confirmed by Lynen's group using the enzyme from rat liver. About the time when allosteric control was first postulated, Vagelos (1962) proposed that the conformational changes causing acetyl CoA carboxylase to aggregate were linked to the activation of the enzyme.

The way biotin participates in carbon dioxide fixation was established in the early 1960s. In 1961 Kaziro and Ochoa using propionyl CoA carboxylase provided evidence for $^{14}CO_2$ binding in an enzyme-biotin complex. With excess propionyl CoA the ^{14}C label moved into a stable position in methyl malonyl CoA. In the same year Lynen found biotin itself could act as a CO_2 acceptor in a fixation reaction catalyzed by ß-methylcrotonyl CoA carboxylase. The labile CO_2 adduct was stabilized by esterification with diazomethane and the dimethyl ester shown to be identical with the chemically synthesized molecule. X-ray analysis of the bis-*p*-bromanilide confirmed the carbon dioxide had been incorporated into the N opposite to the point of attachment of the side chain. Proteolytic digestion and the isolation of biocytin established the biotin was bound to the ε-NH_2 of lysine.

Fatty Acid Synthase

By 1960 it was clear that acetyl CoA provided its two carbon atoms to the ω and ω–1 positions of palmitate. All the other carbon atoms entered via malonyl CoA (Wakil and Ganguly, 1959; Brady et al. 1960). It was also known that ^{3}H-NADPH donated tritium to palmitate. It had been shown too that fatty acid synthesis was very susceptible to inhibition by *p*-hydroxy mercuribenzoate, *N*-ethyl maleimide, and other thiol reagents. If the system was pre-incubated with acetyl CoA, considerable protection was afforded against the mercuribenzoate. In 1961 Lynen and Tada suggested tightly bound acyl-S-enzyme complexes were intermediates in fatty acid synthesis in the yeast system. The malonyl-S-enzyme complex condensed with acyl CoA and the ß-keto-product reduced by NADPH, dehydrated, and reduced again to yield the (acyl+2C)-S-enzyme complex. Lynen and Tada thought the reactions were catalyzed by a multifunctional enzyme system.

Between 1961 and 1963 Vagelos and colleagues examined fatty acid synthesis in *E. coli* and *C. kluyveri*. They confirmed the extreme sensitivity to thiol inhibitors of acetyl CoA transfer to an SH group on the synthase. In the bacterial system this SH was easily distinguished from a less reactive group originating from pantotheine, which was also part of the fatty acid synthase complex. In *E. coli* the pantotheine moiety was eventually isolated as acyl carrier protein (ACP). Acetoacetyl was bound onto the pantotheinyl SH and, when bound, it could be reduced by NADPH to give a butyryl group, still attached to the enzyme complex.

Similar enzymic activity was found by Lynen in his highly purified yeast synthase. He demonstrated the second reduction step using *S*-ß-hydroxyl-*N*-acetylcysteamine which was dehydrated to *S*-crotonyl-*N*-acetylcysteamine and reduced with NADPH to the butyryl derivative. In 1962, Lynen proposed that mammalian and yeast fatty acid synthases comprised soluble multifunctional enzymes which had two essential SH groups: one derived from cysteine to which the initiating acetyl group was bound, and the second from pantotheine onto which the incoming malonyl groups attached. Intermediates remained bound to the synthase throughout the buildup of the fatty acids and did not exchange with potential precursors in the medium. Later work on the structures of the mammalian, yeast, and *E. coli* synthases has confirmed this view, and established the contrast between the mammalian multifunctional, dimeric enzymes and the multienzyme complex in *E. coli* which can be dissociated into its separate, functioning proteins and reassembled.

Lynen's work on the breakdown and synthesis of fatty acids was recognized by the award of a Nobel prize in 1964.

REFERENCES*

Bloor, W.R. (1943). The Biochemistry of the Fatty Acids. Reinhold Publishing Co., New York.

Green, D.E. & Gibson, D.M. (1960). Fatty acid oxidation and synthesis. In: Metabolic Pathways (Greenberg, D.M. Ed.) Vol. I, pp. 301-340. Academic Press, New York.

Lipmann, F. (1948-1949). Biosynthetic Mechanisms. Harvey Lectures XLIV, 99-123.

Vagelos, P.R. (1964). Lipid metabolism. Annu. Rev. Biochem. 33, 139-172.

*Pathways for fatty acid synthesis and breakdown are given in Appendix 2.

Chapter 8

THE IMPACT OF ISOTOPES: 1925-1965

INTRODUCTION

The use of isotopes in biochemistry, particularly radioisotopes, took off after World War II. Developments in electronics and nuclear energy, and the construction of piles in the U.S. and the U.K., enormously improved the production and detection of radioisotopes. At the same time the introduction of paper and ion-exchange chromatography (Chapter 10) revolutionized analytical methods for the separation of low molecular weight compounds, enabling intermediates to be separated rapidly, identified, and estimated. By 1945 strategies for the evaluation of metabolic pathways and cycles were familiar, thanks to the work of Krebs and the pre-war German schools.

Hevesy in 1923 was the first to use an isotope—radium D, a natural isotope of lead—for tracer experiments. By the early 1940s systematic studies of ^{32}P uptake in animals were underway with Hevesy investigating the damaging effects of ionizing irradiation on ^{32}P uptake into DNA in cell tumors (1947). The earliest autoradiographic experiments with biological material were probably those of Lacassagne (1924) who examined the distribution of polonium in biological material.

Applications of isotopes in biochemistry have been manifold. Besides tracing pathways and following the rates of isotope uptake and their intracellular localization, labeling molecules in selected stable positions allowed studies to be made of reaction details, such as those in citrate metabolism (Chapter 5) or squalene biosynthesis (see below). Inserting labels into exchangeable positions gave further insights into mechanisms (e.g. ^{18}O studies by Mildred Cohn). Only a few examples of the use of isotopes will be considered here.

THE DETECTION OF ISOTOPES

Radioisotopes

Although physicists were able to detect and estimate weak ß-ray emitters with elaborate windowless counters, these were not very practical for biochemical experiments; a 24 h working day was required to count 3 to 4 samples! It was not until the late 1940s that isotopes with strong ß emissions could be detected by Geiger-Muller ionization counting. ^{32}P, with a ß-emission which was not stopped by thin glass, was the easiest isotope to use for biochemical experiments. The earliest liquid counters operated at >1000 volts and were quenched by ethanol vapor. In the 1950s halogen-quenched tubes were introduced. These needed lower voltages and had longer useful lives. In Europe, ^{32}P incorporation into plant and animal tissues was studied intensively from the end of World War II (see below).

By the late 1940s methods of ionization counting had been developed for the estimation of ^{14}C. Thin layers of suitably prepared material could be counted in windowless counters, with the emissions being detected by the probe in a flow of inert gas. Otherwise ^{14}C-containing material could be combusted to $^{14}CO_2$ and this drawn over the probe. Neither method was very convenient to use and the instruments were costly. Early experiments with ^{14}C were mainly performed in the U.S. (see below).

Autoradiography offered an alternative technique. The earliest experiments followed the uptake of radioiodine into the thyroid by placing the tissue sections in direct contact with photographic plates (Hamilton, Soley, and Eichorn, 1940; Leblond, 1943). Belanger and Leblond introduced the use of liquid emulsion in 1946. Initially this was painted onto sections mounted on microscope slides. Later, slides were dipped into liquid emulsion (Joftes and Warren, 1955) or wrapped around with stripping film (Doniach and Pelc, 1950). Semiquantitative comparisons could be made of uptakes into different cells or tissues under various conditions. By the late 1950s autoradiographic localization of isotopes, especially ^{3}H, had been extended to ultrastructural studies with electron microscopy (Liquier-Milward). Weak emitters were favored because their shorter ionization tracks allowed more exact localization of the site of uptake.

Tissue or cell autoradiography required considerable expertise. Artifacts, for example from differential pressure over the sections, were very easily introduced but could be recognized from the appearance of the grains. A relatively high humidity was essential to keep the background down. Incorporation could only be followed into "fixed" materials—insoluble macromolecules—and there were conflicting requirements between retaining good cytology and loss of label during the preparation of the slides (see also Chapter 9). Differences in the energies of emission between ^{14}C and ^{3}H made it possible to discriminate between, for example, ^{3}H-labeled DNA and ^{14}C-labeled protein by using two layers of film. Autoradiography became an important complement to cell fractionation for the localization of reactions (Chapter 9).

Liquid scintillants were developed from 1949, based on *p*-terphenyl and diphenyloxazole in a toluene-based solvent. At the beginning there were serious problems from high backgrounds. These were reduced by lowering the temperature and by the introduction of coincidence counting. In the earliest counters corrections had to be made if counting rates were high. The first coincidence times were about 1 μsec but by the early 1970s these had fallen to 20-50 nsec. Automatic sample changing became available in the late 1950s when external standards were also provided to allow more accurate corrections for quenching. The first multi-channel scintillation counters were on the market by 1965. These gave backgrounds of 20-25 cpm, with about 60% efficiency for ^{3}H counting and discrimination so that not more than 10% ^{14}C counts were recorded in the tritium channel.

Stable Isotopes

By 1930 experiences of the earliest workers with radioisotopes, especially α emitters, had offered dramatic evidence of their dangers. It was recognized that stable isotopes were preferable to radioactive ones. Unfortunately quantitative estimation of stable isotopes, except ^2H, required expensive mass spectrometers and considerable time and care in the preparation of samples which would be suitable for assay. Early experiments with ^{13}C, ^{15}, N and ^{18}O were mainly performed in the U.S. Estimation of deuterium was very much easier, from differences in specific gravity between ^2H_2O and ^1H_2O, such as in the falling-drop method used by du Vigneaud.

Rates of reaction of urease with deuterated and especially tritiated urea were markedly reduced compared with the rate with the unlabeled (^1H) substrate, but usually isotope effects are insignificant biochemically except in rigorous kinetic studies.

THE AVAILABILITY OF ISOTOPES FOR BIOCHEMICAL USE

The first isotopes for biochemical use were "spin-offs" following the construction of cyclotrons and the piles at Harwell, Brookhaven, and Oak Ridge. Often compounds had to be synthesized by users from e.g. $^{14}CO_2$ (see below). Even in the mid-1950s it was customary to isolate labeled compounds oneself from crude starting material obtained from *Chlorella* which had been grown in the presence of $^{14}CO_2$. The introduction of catalyzed ^3H exchange reactions to insert ^3H into stable positions in an enormous range of compounds revolutionized isotope experiments, and reduced their costs by an order of magnitude.

THE DYNAMIC STATE OF BODY CONSTITUENTS

In 1933, Schoenheimer, who was medically qualified and had been working with Aschoff in the Pathology Institute in Freiburg, moved to Columbia University, New York, and was joined the next year by David Rittenberg. Rittenberg had just spent some time in Urey's laboratory in the Rockefeller Institute learning techniques for handling deuterium. Their first experiments concerned the metabolism of deuterated fatty acids in rats and the demonstration (see below) that ^2H from heavy water was incorporated by the animals into fatty acids and cholesterol.

With Sarah Ratner they next studied the uptake of ^{15}N-labeled amino acids into tissue proteins. Glycine and L-leucine, also carrying ^{2}H on its alkyl carbon atoms, were fed to rats for three days. The amino acids made up about 7% of the dietary intake of nitrogen, 56.5% of the ^{15}N from leucine and 44.3% of that from glycine were incorporated into protein. Serum proteins and those from liver, intestinal wall, and kidney were heavily labeled; proteins from skin and hemoglobin contained little excess ^{15}N. With the doubly labeled leucine both ^{2}H and ^{15}N were retained in leucine in the proteins. ^{15}N was also incorporated into all the other amino acids except lysine, but principally into glutamate and aspartate, as would be expected from transamination and the action of L- glutamate dehydrogenase (Chapter 6). Schoenheimer and Rittenberg concluded "peptide bonds [are] essential parts of the proteins, and one may conclude that they are rapidly and continually opened and closed in proteins in normal animals. The experiments give no direct indication as to whether the rupture [of the peptide bonds] is complete or partial."

If ^{15}N ammonium citrate was administered, and glutamate, aspartate, and glycine isolated from liver and intestinal wall protein, all showed ^{15}N uptake. From the results of labeling studies, Schoenheimer finished his Edward K.Dunham lectures in Harvard in 1941 with the phrase—"the structural materials [of the body] are in a steady state of flux. The classical picture must thus be replaced by one which takes account of the dynamic state of body structure"—an idea which has become an integral part of biochemistry since that time, and which was almost totally dependent on the introduction of isotopes for its discovery.

STUDIES WITH DEUTERIUM

The Biosynthesis of Methionine; Transmethylation

Creatine was first isolated in 1835 by Chevreul; 20 years later Dessaignes showed it to contain a methyl group. Choline was obtained from lecithin in bile by Strecker in 1849 and methionine isolated by Mueller in 1922. That methionine contained a methyl group linked to sulfur was demonstrated by Barger and Coyne in 1928.

The first indication that methylation occurred *in vivo* was probably the observation that if pyridine was given to animals, methyl pyridine was excreted in their urine (His, 1887). From his studies on the fate of administered tellurides and their excretion as methyl telluride, Hofmeister (1894) proposed methylation might be a normal metabolic process in animals. This was conclusively established as a result of extensive work from du Vigneaud and his group between the late 1930s and 1955. Standard nutritional experiments with rats were combined with pioneering studies on the metabolism of ^{2}H-labeled compounds (see du Vigneaud, 1952).

In 1931, Jackson and Bloch reported that methionine supplemented growth in rats which were being kept on a low cystine diet. Du Vigneaud showed that homocystine could replace methionine, and proposed that both methionine and homocysteine were metabolized *in vivo* to give cysteine. He next synthesized L-cystathionine and showed it too could replace cysteine in a diet containing minimal methionine. Stetten, using ^{15}N, found the nitrogen of cysteine originated from serine. Du Vigneaud, who was an excellent chemist, then synthesized doubly labeled methionine, $CH_3{}^{34}S{\cdot}{}^{13}CH_2{\cdot}{}^{13}CH_2{\cdot}CHNH_2{\cdot}CO_2H$, and fed this to two rats which had been shaved before the start of the experiment. After 35 days the hair was clipped again, methionine and cystine were isolated from keratin, and the ^{34}S and ^{13}C contents determined—the carbon as carbon dioxide and the sulfur as hydrogen sulfide after combustion of the amino acids in a stream of hydrogen. No ^{13}C from methionine was incorporated into cysteine, which however contained approximately the expected level of ^{34}S. When the experiment was repeated with racemically resolved L-^{35}S-cystathionine, this was conclusively shown to be the precursor of cystine, so confirming that cysteine could be synthesized *in vivo* from homocysteine and serine.

The question therefore arose about the fate of the methyl group from methionine. When minimal amounts of methionine were used to supplement the diet of rats given homocysteine as their main source of sulfur, the rats did not usually thrive, and at death had fatty accumulations in their livers. Best and his co-workers had earlier reported the efficacy of choline as a lipotropic agent, facilitating the mobilization of fat from the liver. Du Vigneaud therefore tried supplementing homcysteine in the rats' diet with choline and found methionine was no longer needed. Betaine, the oxidation product of choline (Bernhein and Bernhein, 1937; Mann and Quastel, 1937) was also effective.

Methionine was next synthesized from C^2H_3I and homocysteine, the deuterated methyl iodide having been prepared from C^2H_3OH obtained after the reduction of carbon monoxide by $^2H_2$ under pressure. Rats were fed the labeled methionine; after 3 weeks 57.4% of the choline was derived from the methyl group of methionine and after 94 days, 88.6%. Du Vigneaud therefore concluded transmethylation took place from methionine without significant breakdown of the methyl group. Other methylated compounds like creatine were also shown to have received their methyl groups from methionine; the methyl on creatine was not further transferred. Net synthesis of creatine following transmethylation from methionine to guanidoacetic acid was reported by Borsook and Dubnoff (1940-1947). Transmethylation is not restricted to higher animals. From 1935, Challenger and his associates studied methyl group transfer from the thetins (sulfur analogues of betaines) in fungi and algae.

Mechanism of Transmethylation

The mechanism of transmethylation was then examined. A series of deuterium-labeled methylated compounds were synthesized by du Vigneaud's group, including arsenocholine, trimethylamine, dimethylglycine, and dimethylthetin. Of these only betaine and dimethylthetin served as methyl donors. In 1949 Dubnoff found that choline could only act as a donor under aerobic conditions, when it was oxidized to betaine.

The final step in our understanding of transmethylation followed from the observation by Cantoni (1951) that betaine and dimethylthetin only acted as methyl donors in the presence of homocysteine, i.e. after the methyl group had been transferred to give methionine. Methionine would only transmethylate if ATP was available. *S*-adenosyl methionine was therefore proposed as the primary methyl donor, a suggestion confirmed after the compound had been synthesized by Baddiley and Jamieson in 1954.

One further very significant result from these experiments came from noting that, when fed the standard diet of homocysteine without any added methyl donor, the occasional rat recovered its growth rate. Du Vigneaud realized this might be due to microflora in the animal's gut synthesizing methyl groups *de novo*, an idea substantiated when Toennies, Bennett, and Medes (1943-1944) showed growth was always prevented if the rats were fed sulphasuccidine, a bacteriostatic agent. Notre Dame Research Institute maintained a colony of germ-free rats. Du Vigneaud therefore collaborated with the Institute, examining the extent of ^{2}H incorporation from labeled drinking water into choline. This was found only if the germ-free rats were given folic acid and vitamin B_{12}, so establishing that in the presence of these essential vitamins limited synthesis of methyl groups can occur in animals. It was serendipitous that the diets used by du Vigneaud for most of his experiments contained only minimal amounts of folic acid and B_{12}, so that little synthesis of endogenous methyl groups could occur.

^{14}C-ACETATE AND CHOLESTEROL BIOSYNTHESIS

Analyses of the biosynthesis of methionine and of transmethylation were based on comparatively straightforward experiments using suitably labeled compounds in conjunction with classical nutrition studies. The work was virtually complete within 15 years. Determining the pathway of cholesterol biosynthesis was much more difficult, and extended from the early observation of Rittenberg and Schoenheimer (1937) that mice given heavy water to drink, incorporated deuterium into their fatty acids and cholesterol, to continuing work on the regulation of cholesterol metabolism and its relation to human disease. Tracing the path by which carbon atoms from acetate were incorporated into cholesterol (Bloch and Rittenberg, 1946) was achieved significantly before the enzymology of the individual steps became clear. The work required the development of special chemical degradative procedures so that the origins of the different carbon atoms in cholesterol could be unequivocally assigned. High resolution mass spectrometry with selectively labeled ^{2}H intermediates was necessary to establish the stereochemistry of the condensation of the isoprene units. The previously unknown enzymes had to be isolated and characterized.

The first experiments demonstrating acetate incorporation into cholesterol were followed by quantitative experiments showing that the ratio of methyl carbon: carboxyl carbon from acetate was 1.27:1 (Little and Bloch, 1950). Bloch's group and Brady and Gurin then tried to identify possible early products from acetate condensation. Rats were given a range of ^{14}C-labeled compounds—acetone, acetoacetate, pyruvate, butyrate, octanoate and the isopropyl group of isovalerate—all of which showed the ^{14}C label in their cholesterol. Only after the isolation of acetyl CoA and the elucidation of the fatty acid spiral by Lynen did it become clear that interpretation of these tracer experiments was doomed to ambiguity because of the ease of their breakdown to or resynthesis from acetate.

Squalene had been isolated from shark's liver by Channon in 1929. He, Heilbron, and Robinson all postulated it was a potential precursor of cholesterol. In 1952, Bloch and his colleagues established the differential origins of the carbon atoms in the cholesterol side chain from the methyl or carboxyl carbon atom of acetate:

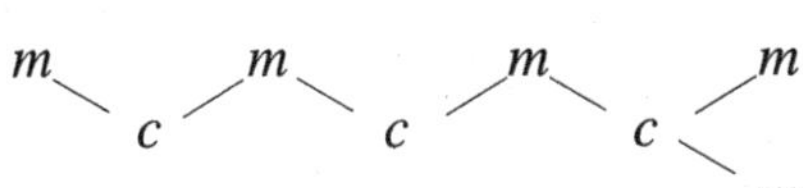

Bloch therefore suggested isoprene units could be condensed first to give squalene and then cholesterol, an extension of Ruzicka's isoprene rule for the biosynthesis of linear and cyclic terpenoids. It was then necessary to show that selectively labeled ^{14}C-acetate could get incorporated into squalene with the correct distribution of ^{14}C and that this squalene could give rise to cholesterol, also with the appropriate position of the ^{14}C label.

Between 1953 and 1957 Cornforth and Popjak in the U.K. and Bloch in the U.S. unequivocally established the origins of all the carbon atoms in cholesterol, 15 being derived from the methyl group of acetate and 12 from its carboxyl group. This ratio of 1.25 was very close to the value of 1.27 reported earlier by Bloch. Langdon and Bloch then (1953) showed ^{14}C squalene was incorporated into cholesterol and in the same year Woodward and Bloch put all the results together to suggest a route for

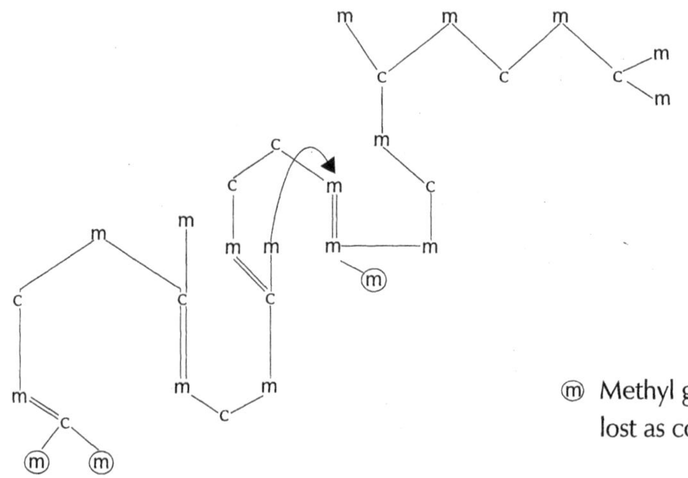

the cyclization of the linear 30C molecule and the elimination of the extra 3C atoms which was consistent with the isoprene rule:

Analysis of the details of the pathway was helped by the discovery by Nancy Bucher (1953) that cholesterol synthesis took place in cell-free post-mitochondrial supernatants. ATP, Mg^{2+} and NAD^{+} were required. Tchen and Bloch extended these findings to show that squalene could be formed anaerobically but the conversion of squalene to cholesterol was oxygen dependent, the oxygen of the intermediate lanosterol being derived from $^{18}O_2$ not $H_2{}^{18}O$. It therefore became possible to focus either on the conversion of acetate to squalene or on the latter's cyclization to the sterol.

The route for the cyclization was easier to determine than the identification of the very reactive isoprene unit, and was understood in outline by 1960. Studies of labeled compounds detected within 10 min. of ^{14}C-acetate addition to intestinal preparations showed label in squalene, lanosterol, and a further, unidentified ring compound, all with higher specific activities than cholesterol. By 75 min cholesterol was the main labeled compound. Clayton and Bloch then confirmed that lanosterol, previously known from sheep's wool, was converted to cholesterol with the extra three (methyl) carbon atoms being lost as carbon dioxide.

Identification of possible intermediates in the cyclization was attempted by adapting the isotope dilution method which had been devised by Rittenberg for the estimation of amino acids in proteins. Potential intermediates were added as carriers to cholesterol-synthesising systems; after incubation with ^{14}C-acetate the carrier was reisolated and checked for radioactivity. If positive, it was added to a fresh synthesizing system and the formation of labeled cholesterol examined. In this way zymosterol and desmosterol were also shown to be possible intermediates. With such methods it was difficult to ensure that the addition of the carrier was not altering the course of the reactions. Isolation of the relevant enzymes, however, confirmed the provisional route.

Insight into the early stages of acetate condensation came from experiments unrelated to sterol biosynthesis. Tavormina, Gibbs, and Huff (1956) were working with a strain of *Lactobacillus* which required acetate for growth. They found the 6C compound mevalonic acid (MVA) could replace acetate. They then tested DL-mevalonic acid(-2-^{14}C) as a possible precursor for cholesterol and found that in rat liver 43.4% was very rapidly converted into the sterol. If 1-^{14}C MVA was used all the radioactivity was released as $^{14}CO_2$. 2-^{14}C MVA could give an asymmetrically labeled isoprene unit; if C2 condensed with C5 of a second molecule it was possible to predict the labeling expected in cholesterol (C1,7,15,22, and either 26 or 27)—the isotope distribution found experimentally by Isler and colleagues. The suggestion was therefore made that MVA originated from the reduction of ß-methyl,ß-hydroxy-glutaryl coenzyme A (HMGCoA) and that this reduction might be a limiting step in cholesterol biosynthesis. Rudney in 1957 described a thiolase which catalyzed the condensation of two molecules of acetyl CoA to give acetoacetyl CoA. HMGCoA synthase was then isolated and HMGCoA reductase discovered (J.W. Porter et al., 1960s).

Earlier work by Nancy Bucher showed an ATP requirement for cholesterol biosynthesis. The involvement of phosphorylated intermediates was established by Cornforth, Popjak, and their associates in the early 1960s with the discovery of kinases which successively phosphorylated MVA to MVA-P and MVA-P to MVA-PP. MVA-PP was decarboxylated and dehydrated to give the biological C5 isoprene unit, isopentenyl pyrophosphate. This undergoes successive head-to-tail condensations to give the linear 15C terpenoid, farnesyl pyrophosphate.

Back in 1951, Vennesland and Westheimer had used deuterium labeling to show there were two classes of dehydrogenase donating hydrogen atoms to the pyridine ring in NAD^+ or $NADP^+$, the α and ß reductases. Cornforth established α–NADH had the *R* and ß the *S* configuration. He and Popjak then used ^2H-labeled MVA or farnesyl PP to establish the precise stereochemistry of the condensation of the 15C units. With 5-dideutero MVA as precursor, squalene was isolated and degraded so that the central four carbon atoms formed succinic acid. Analysis by mass spectrometry indicated 80% of the molecule was trideuterated, i.e. one of the four hydrogens on the central two carbons of squalene was lost during the condensation of the two farnesyl molecules. As the stereochemistry of H transfer to NAD^+ was known the absolute configuration of the trideuterosuccinic acid was established. That the centre of the squalene molecule was asymmetric with one C carrying both ^1H and ^2H was confirmed with 1-dideuterofarnesyl PP-2-^{14}C as precursor. For their work in the field of cholesterol biosynthesis Bloch and Cornforth both received Nobel prizes.

STUDIES WITH ^{32}P

Between 1945 and 1965, biochemical studies with ^{32}P fell into two classes—investigations which uncovered fundamental aspects of phosphorus metabolism, and a large number of experiments comparing ^{32}P distribution between normal and experimental systems. One example of the latter has already been mentioned—Hevesy's observations on the radiosensitivity of ^{32}P uptake into DNA. These comparative studies, many of which were highly significant, will not be considered here. Hevesy's work with ^{32}P also identified one of the foundations of biochemistry, the metabolic stability of DNA in resting cells. His work, that of J.N. Davidson's group in Glasgow, and others, established the great rapidity with which ^{32}P was taken up by RNA in metabolically active cells.

W.C. Schneider and Schmidt and Thannhauser (1945-1950) introduced procedures for the fractionation of ^{32}P-labeled compounds in tissues. Inorganic phosphate is very tightly bound to proteins. Both procedures therefore recommended extensive washing in cold dilute acid, to which unlabeled P_i or ATP might be added, to reduce $^{32}P_i$ contamination in later extracts. The rapidity with which acid-labile phosphates turned over, especially in liver, was an important early finding. By the early 1950s various chromatographic methods were used to separate the different acid-soluble nucleoside mono-di-and tri-phosphates from each other and from P_i so that the movement of ^{32}P through the soluble nucleotide pool could be followed.

Treatment with hot organic solvents was the next step in the tissue fractionation, to remove lipid-phosphorous and breakdown lipid-protein interactions. In the Schneider procedure, nucleic acids were then extracted in hot dilute trichloroacetic or perchloric acid, leaving a protein residue with any phosphoprotein links still intact. This method was to become particularly useful when ^{3}H thymidine became the preferred label for DNA in the early 1960s. For investigations where both RNA and DNA were to be examined the Schmidt-Thannhauser process was often chosen. Here the lipid-extracted material was hydrolyzed with dilute sodium hydroxide releasing RNA nucleotides and any hydroxyamino acid bound phosphorus. DNA could be precipitated from the extract but the presence in the alkaline hydrolysate of the highly labeled phosphate released from phosphoprotein complicated the analysis of ^{32}P uptake into RNA. In the late 1950s extraction with phenol was introduced allowing more rigorous separation of DNA and RNA and much less contamination from protein.

In 1949 Marshak and Calvet and Barnum and Huseby showed nuclear RNA had a higher turnover of ^{32}P than cytoplasmic RNA. Ten years later when methods of isolating nuclei and nucleoli were available, different fractions of nuclear RNA with different compositions and turnovers were described. These included Hn-RNA—material which had a relatively high molecular weight—was polydispersed on sucrose gradients, differed in composition from rRNA, and was rapidly labeled. mRNA had been detected in *E. coli* by Volkin and Astrachan in 1956. Hn-RNA was thought to be related to mRNA. Kinetic analysis with ^{14}C-adenine and ^{14}C-adenosine in Hela cells (H. Harris and Watts, 1959-1962) indicated that a large fraction of nuclear RNA was degraded to acid-soluble nucleotides before leaving the nucleus.

By the early 1960s it was clear that simple experiments on ^{32}P uptake were inadequate to distinguish different species of RNA in mixed populations of cells at different stages of the cell cycle. The future for labeling experiments with ^{32}P in cell biology was to lie mainly with synchronized cells and for studies on nucleic acids, with molecular biological techniques.

Analyses with ^{32}P as a label had however already made an important contribution to cell biology with the identification of the precise period in the eukaryotic cell cycle when DNA was synthesized. In 1951, Howard and Pelc followed the incorporation of ^{32}P into meristematic cells of *Vicia faba* autoradiographically. Roots were exposed to 0.2 μCi $^{32}P_i$/ml for 24 h. Control experiments indicated that the level of isotope employed produced no deleterious effect on subsequent growth. Treatment with DNAase was used to confirm that the grains originated from ^{32}P uptake into DNA. The autoradiographs showed clearly the isotope was incorporated into DNA in cells in interphase and not in prophase nor any other stage of mitosis. Howard and Pelc reported: "DNA is not synthesized during cell division but only during interphases. [T] here is a time lag between the end of synthesis and the beginning of visible prophase of about 6 h." These experiments gave rise to current terminology for the phases in the cell cycle: G1, S, G2, and M (Interphase Gap 1, DNA Synthesis, Interphase Gap 2, and Mitosis).

Three years later, Lajtha, Oliver, and Ellis performed similar studies with human bone-marrow cultures exposed to $^{32}P_i$ or ^{14}C-adenine. Control smears were treated with M HCl at 60 °C for 6.5 min to remove ^{32}P not incorporated into DNA. Grain counts were made over individual nuclei so that the rate of uptake into DNA could be estimated. The cycle time for the dividing cells in the culture was 40-48 h. DNA synthesis took 12-15 h in the second half of the cycle and was divided from mitosis by a 3-4 h non-synthesizing period (G2).

Isotopic experiments on the cell cycle could also be performed with ordinary (non-autoradiographic) biochemical procedures if synchronized cells were available. The earliest animal preparation providing such a system for use *in vivo* was the regenerating rat liver. Stowell (1949) and Abercrombie and Harkness (1951) noted that DNA synthesis in livers from partially hepatectomized rats occurred about 24 h after operation, whereas mitoses were not seen until about 30 h. This was confirmed by Kelly (1953) and Holmes (1956) who were also involved in studies on the biochemical effects of ionizing irradiation. Kelly and Holmes found DNA synthesis was much more sensitive to exposure if the animals were irradiated at the time of injury (when the stimulus to liver regeneration was given) rather than in S phase, an observation from which many later experiments originated (for refs. see Ord and Stocken, 1980).

THE CALVIN CYCLE

Another of the fundamental biochemical studies in the early 1950s linked with the use of isotopes was the discovery by Calvin and his associates at Berkeley of the pathway of carbon dioxide fixation in higher plants. The experiments were elegant in their simplicity and definition. Illuminated photosynthetic algae (*Scenedesmus* or *Chlorella*) were exposed to $^{14}CO_2$ for periods from 10 s to a few minutes. The reactions were rapidly stopped by quenching the cells in hot alcohol and the 3C and 6C compounds in the alcoholic extract then examined by two-dimensional paper chromatography and autoradiography. After 10 s, phosphoglyceric acid (PGA) was the main labeled compound. By 60 s, ^{14}C activity was present in many compounds but especially ribulose diphosphate and 5,6 and 7C sugar phosphates. The distribution of ^{14}C in PGA and the 6C sugar after 15 s showed 49% of the ^{14}C in the CO_2H of PGA and 25% in each of the other two carbon atoms. In the hexose, C3 and C4 were equally labeled, together having half the total radioactivity of the molecule. C2 and C5 and C1 and C6 shared the remaining ^{14}C, i.e. the 6C sugar had exactly the distribution of ^{14}C which would be expected from the condensation of two molecules of PGA to give F1,6diP by a reversal of the glycolytic pathway and aldolase reaction. The detailed distribution of ^{14}C was then examined in a number of compounds to identify which molecule was the acceptor for the $^{14}CO_2$ and gave rise to the labeled PGA. C3 of ribulose diphosphate (RuDP) was very highly labeled.

That RuDP was the carbon dioxide acceptor was established in further experiments when the light and dark reactions were separated.

The algae were exposed to $^{14}CO_2$ for 5-10 min in the light. Labeling of PGA, RuDP, and hexose monophosphates had by then reached a plateau. When the light was turned off, ^{14}C activity in RuDP declined very rapidly but that in PGA continued to increase for 2 or 3 min. Such a precursor-product relationship identified RuDP as the carbon dioxide acceptor. The reactions linking the other members of the pentose-phosphate pathway were also determined, and the route by which the 7C sugar, sedoheptulose, could be formed and recycled to 3C and 6C compounds by the new enzymes, transketolase and transaldolase. The enzyme responsible for carbon dioxide fixation, then called carboxydismutase and now ribulose bisphosphate carboxylase (RUBISCO), was isolated from algae and higher plants.

Light was required for the generation of NADPH needed to reduce phosphoglyceric acid to give glyceraldehyde 3-phosphate. The NADPH was then used to give the three molecules of ATP needed to phosphorylate ribulose phosphate and the triose phosphates, so completing the Calvin cycle.

Many seminal experiments with isotopes have not been recounted. The classical experiments of Hershey and Chase in 1952 demonstrated information transfer from phage DNA to its host *E. coli*. Taylor, Woods, and Hughes (1957) examined metaphase chromosomes in root cells of the lily *Bellavalia romana* autoradiographically and showed DNA replication to be semi-conservative. Meselson and Stahl in 1958 used density gradient centrifugation to prove DNA replication in *E.coli* was also semi-conservative. These experiments were to become the cornerstones of molecular biology. The advent of isotopic tracers heralded the completion of that stage in biochemistry associated with the establishment of pathways in intermediary metabolism. Not that all pathways had been discovered by the early 1960s, but the techniques necessary for the elucidation of reaction sequences were now clear.

REFERENCES

Calvin, M. (1955). The Photosynthetic Carbon Cycle. In: Proceedings of the 3rd International Congress of Biochemistry (1956) (Liebecq, C. Ed.) pp. 211-227. Vaillant-Carmanne, Liege.

Cantoni, G.L. (1955). Enzymatic Mechanisms and the Biological Significance of Transmethylation Reactions, pp. 233-237, ibid.

Crook, M.A., Johnson, P., & Scales, B., Eds. (1972). Liquid Scintillation Counting. Heyden & Son, London.

Davidson, J.N. (1950). The Biochemistry of the Nucleic Acids. Methuen.

Harris, H. (1968). Nucleus and Cytoplasm. Clarendon, Oxford.

Hevesy, G. (1947). Radioactive indicators in turnover studies. Advances in Enzmol. 7, 111-214.

Howard, A. & Pelc, S.R. (1951). Nuclear incorporation of ^{32}P as demonstrated by autoradiographs. Exptl. Cell Res. 2, 178-187.

Lajtha, L.G., Oliver, R., & Ellis, F. (1954). Incorporation of ^{32}P and ^{14}C adenine into DNA by human bone-marrow cells *in vitro*. Brit. J. Cancer 8, 367-399.

Ord, M.G. & Stocken, L.A. (1980). Nucleoprotein changes in the cell cycle: Interphase studies with regenerating rat liver. Biochem. Soc. Trans. 8, 759-766.

Overman, R.T. & Clark, H.M. (1960). Radioisotope Techniques. McGraw Hill, London.

Popjak, G. (1958). Biosynthesis of cholesterol and related substances. Annu. Rev. Biochem. 27, 533-560.

Rogers, A.W. (1977). Techniques of Autoradiography, 3rd ed. Elsevier, North Holland, Amsterdam.

Schoenheimer, R. (1942). The Dynamic State of Body Constituents. (Republished 1964). Hafner Publishing, New York.

Vigneaud, V. du (1952). A Trail of Research. Cornell University Press.

Wolstenholme, G.E.W., Ed. (1951). Isotopes in Biochemistry. CIBA Foundation Symposium. J. & A. Churchill, London.

Wolstenholme, G.E.W. & O'Connor, C.M., Eds. (1956) Ionizing Radiations and Cell Metabolism. CIBA Foundation Symposium. J. & A. Churchill, London.

Chapter 9

BIOCHEMISTRY AND THE CELL

THE AGE OF CLASSICAL MICROSCOPY: CA. 1840-1940

The fall and resurrection of the cell theory were reviewed in Chapter 2. Under the microscope the nucleus of the cell was its most prominent feature; its boundary was thought to be delineated by a membrane with the space between filled by a homogeneous transparent fluid which might contain granules (Schwann, 1839). Thomas Huxley in 1850 described the cell as a nucleus and protoplasm surrounded by a membrane (periplast), although the existence of a membrane chemically distinguished from the protoplasm was not universally accepted. The word "protoplasm" is thought to have been used first by Purkinje (1839) to describe matter possessing a molecular constitution permitting it to manifest life—a definition posing the question "which cell constituents were alive? By 1850 Ferdinand Cohn had recognized that the protoplasm most commonly observed in plant cells must have properties similar to the contractile substance, sarcode, in animal cells. Protoplasm was the primary location of "life as action."

Virchow, who was largely responsible for the acceptance of the cell theory, developed microscopy of cells from normal and diseased tissues as a major tool (histopathology) in the clinical armory. He believed the vital functions of the cell, growth, maintenance, and multiplication were discharged by its nucleus; the specialised, distinguishing functions were made possible by the extranuclear constituents. In a Sunday evening lecture in Edinburgh in 1868, "On the Physical Basis of Life", Thomas Huxley described cells as protoplasmic masses usually furnished with nuclei. Cells were the constitutive elements of plants and animals. Protoplasm was semifluid, contractile (vital), albumoid, coagulable (protein), and self-replicatory—the physical basis of life.

Problems over the definition of "protoplasm" became increasingly acute as more intracellular inclusions became visible. Kollicker in the middle of the nineteenth century described granules in muscles (sarcosomes); components of the mitotic apparatus were also seen as were fat droplets, starch granules, and pigmented bodies. The Golgi apparatus was described in 1898 and mitochondria in 1903 (Benda). Many of these bodies multiplied and divided apparently synchronously with the division of the cell. Pfluger considered protoplasm to be a giant molecule whose side chains were involved in assimilation and oxidation. Fermentation was a function of protoplasm as a whole. In 1875 he defined life as "intracellular warmth" which was produced by oxidative disruption of "giant albumen molecules" which could multiply through polymerization. Ideas about protein structure were still constrained by the colloid theory (Chapter 10).

By the early years of this century the cell was generally recognized as the smallest unit capable of independent life. Gowland Hopkins in 1913 first clearly formulated ideas which would be the death of "protoplasm." "Life is the expression of a particular dynamic equilibrium which obtains in polyphasic systems ... life is a property of the cell as a whole."

TECHNIQUES IN VISIBLE MICROSCOPY

While van Leeuwenhoek in the seventeenth century had ground lenses so that he could observe and correctly describe various types of bacteria—spirochaetes, cocci, and bacilli—other investigators in the 18th century had difficulties in making lenses, thus leading to irreproducible and fanciful descriptions such as the homunculus in sperm. J.J. Lister corrected chromatic and spherical aberrations in lenses by 1830 and by the middle of the century achromatic compound microscopes were being manufactured by Zeiss. Numerous dyes were developed by the German chemical industry. Using these, some internal cell structures could be effectively distinguished. Condensers were introduced by Abbe in 1872 to improve light focusing onto the object. With oil immersion to increase the refractive index of the medium surrounding the specimen, up to 1000-fold magnification could be achieved.

To obtain tissue preparations whose constituents were maintained as closely as possible to their state *in vivo*, the material had to be "fixed," i.e. the enzymes inactivated so that cell structures were instantaneously preserved, an almost unattainable ideal. Formalin was the favored fixative, but others (e.g. picric acid), were also employed. Different methods of fixation caused sections to have different appearances. Further artifacts were introduced because of the need to dehydrate the preparations so that they could be stained by dyes, many of which were lipid-soluble organic molecules. Paraffin wax was used to impregnate the fixed, dehydrated material. The block of tissue was then sectioned, originally by hand with a cut-throat razor, and later by a mechanical microtome. The sections were stained and mounted in balsam for examination. Hematoxylin (basophilic) and eosin (acidophilic) (H and E staining) were the commonest stains, giving blue nuclei and pink cytoplasm. Eosinophils in the blood were recognized in this way.

The basis of staining was often uncertain and its extent highly dependent on the precise procedure followed. Quantitative comparisons between different tissues were impossible. In 1829, Raspail used iodine which stains starch blue, and by the 1920s and 1930s cytologists were developing more stains whose chemical interactions were understood. The simplest of these were stains for lipids, such as Sudan III or IV. This was followed in the 1940s by the periodic acid-Schiff reaction (Hotchkiss, MacManus) (PAS staining) which stained glycogen and glycosylated proteins. Of great importance was the introduction in 1924 by Feulgen and Rossenbeck of a procedure unique for DNA. Treatment with M HCl at 60 °C dissociated the acid-labile purine residues from DNA, releasing aldehyde groups which reacted with Schiff's reagent so that DNA stained purple against a greenish colored cytoplasm. RNA, where the purine linkage is acid-stable, did not react. Feulgen staining enabled nuclei to be clearly identified in animal and plant cells. The next step was to find a quantitative method for the estimation of the DNA in the nucleus.

A suitable method for this was introduced by Caspersson (mid-1920s–ca. 1940) who designed and successfully exploited UV microscopy so that the extent of the absorbtion could be determined quantitatively. Selective hydrolysis by RNAase or DNAase was used so that the DNA content of the cell could be estimated. The procedure still required reproducible preparation of sections to allow light to be transmitted and the enzymes to get access to the nucleic acids.

UV microscopes were sophisticated to use and very expensive. A number of workers, notably Pollister and Swift, devised cytophotometric procedures with visible light, using Feulgen staining calibrated against UV microphotometry. The DNA content of the nucleus could then be measured relative to that in other cells. Lymphocytes were used as a standard.

In the 1940s Brachet developed Unna's methyl green-pyronin staining method. Using carefully controlled pH the basic dyes combined selectively with DNA (methyl green) or RNA (pyronin). Such differential staining confirmed the presence of DNA in nuclei and RNA in the cytoplasm and indicated an association between high RNA contents in cells and extensive protein synthesis, thus substantiating Caspersson's reports of high concentrations of RNA in nucleoli.

A different application of visible microscopy was pioneered by Gomori. In 1941 he showed that alkaline phosphatase could be specifically located by its hydrolysis of soluble phosphate esters (initially glycerophosphate). If calcium ions were present in the medium in which the sections were incubated, insoluble calcium phosphate precipitated as a result of the action of the hydrolase. The site of the precipitate could be visualized if cobalt or lead salts were subsequently added to replace calcium and the sections exposed to hydrogen sulfide. In principle many hydrolases and other enzymes could be studied using the appropriate substrates and precipitants. It was important to ensure that the products of the enzyme reactions did not diffuse from the sites where the enzymes were located. It was also essential that the reagents could reach the enzyme site.

Unfortunately there were serious problems with the procedure. A major difficulty was the compromise between adequate fixation so that cell structures were protected from autolysis, but sufficient enzyme activity preserved for the hydrolysis to occur. Many hydrolases are robust and very active; partial inactivation still allowed sufficient activity for the reaction to be detected. An improvement introduced in the 1950s was to section tissues in the frozen state after they had been dropped into cooled isopentane and dehydrated *in vacuo* or in absolute alcohol at -60 to -70 °C. Many of the best results were obtained with enzymes located at cell surfaces. Semiquantitative comparisons between tissues were reported.

A procedure which could be applied to living cells was vital staining. Ehrlich in the 1890s had found methylene blue and neutral red were taken into cells. The dyes did not penetrate the nucleus but otherwise transparent cells could now be visualized. Conversely Trypan Blue did not penetrate healthy cells but if the cells were damaged the dye entered. Janus Green was found by Cowdrey (1918) to penetrate cells and stain the large granules (mitochondria)—evidence that in cells *in vivo* mitochondria were the sites of oxygen uptake.

Another important development of visible microscopy which became available in the 1950s was interference microscopy. In the simpler phase contrast microscopy, when phase selected light passed through a transparent object, differences in refractive index showed up as a difference in light intensity. With interference microscopy, which was more expensive but had fewer aberrations, differences in refractive index appeared as differences in color. Both these methods could be used with living, unstained cells so avoiding all fixation and dehydration artifacts. When linked to time-lapse cinematography, either procedure could be used to show movements of and in living cells, their responses to environmental changes, and their behavior during the cell cycle. In spite of the distortions of time-lapse, watching living cells is an important correction to the static view of cells given by classical or electron microscopy.

UNVEILING CELL ULTRASTRUCTURE

Classical microscopy was largely the occupation of professional cytologists. Some biochemists were excited by the possibilities of cytochemical staining for enzymes and the Feulgen technique for DNA, but because of the technical difficulties most biochemists, with their training in quantitative analyses, were more confident working with soluble enzymes or tissue homogenates in intermediary metabolism or in the emerging field of enzymology. Many pathways in metabolism were determined without reference to cell structure.

Between 1948 and 1958 the position changed. The introduction of electron microscopy and its revelations of cell ultrastructure made by a fairly small group of specialist electron microscopists was complemented by the development of differential centrifugation. This allowed the intracellular organelles whose structures were being described by the microscopists to be isolated and their functions examined biochemically. The high-speed centrifuges required for cell fractionation soon became relatively inexpensive. By the mid-1950s differential centrifugation had become a standard procedure. Normally the identity and purity of the cell fraction was checked enzymically and in the electron microscope (EM).

The Introduction of the Electron Microscope: 1930-1960

Electron microscopes were built largely as a result of the work of Knoll and Ruska. By 1931 they had proposed that images of specimens should be obtained if the sample was exposed to an electron beam which could be focused magnetically. In 1933, calculations indicated a resolution of 2.2 Å (0.22 nm) should be attainable, substantially better than with visible light (200 nm), a target finally achieved in the early 1970s. Of major importance was the design of magnetic lenses to collimate the beam of electrons. The first pictures of a biological specimen were reported by Marton in 1934 (a section through a sundew leaf) and by 1938 Siemens, the German engineering firm, offered designs for electron microscopes which became available commercially.

Preparation of biological material for electron microscopy still required fixation, dehydration, and ultrathin sections. Araldite and other resins were used in place of paraffin wax for blocking. At first, specially sharpened steel knives were employed to cut the sections, but from 1950 glass or diamond knives were used which could cut slices 100-200 nm thick. By 1952, Palade and others were obtaining sections cut to 20 nm, giving resolutions of 30 Å (3 nm). As with classical microscopy, the preparations had to be treated to bring out electron opaque contrasts. Initially buffered isosmolar osmium tetroxide was the commonest "stain." Osmium tetroxide was thought to react with unsaturated fatty acid residues, but its penetration was poor, reinforcing the importance of thin sections. For his work on mitochondria in the late 1950s, Fernandez-Moran (Chapter 5) used "negative staining." Fixation by buffered phosphotungstic acid highlighted proteins on the surfaces of cell organelles rather than lipid.

Differential Centrifugation

Differential centrifugation was first introduced by Bensley and Hoerr in 1934 who obtained a large granule fraction containing nuclei and mitochondria. It soon became customary to use a low speed "spin" (10 min at 600 g) to sediment out unbroken cells, contaminating erythrocytes (perfused tissues were often used to diminish contamination from blood), and nuclei. Claude (1943-1946) also recommended that the tissue should initially be passed through a metal sieve or screen to remove connective tissue, so making homogenization easier and less damaging to the intracellular organelles. Claude obtained three fractions: the large granules (mitochondria), small granules (microsomes), and a non-sedimentable fraction.

The choice of medium was important. The presence of ions commonly used in physiological solutions was undesirable because the particles aggregated. Sucrose was introduced as a cheap, highly purified, nontoxic, soluble material which gave biochemically active fractions. Originally 0.88 M sucrose was used because mitochondrial morphology in the EM was then excellent. The mitochondria were elongated and rod-shaped. However, if the sucrose was isosmolar (0.34 or more usually 0.25 M), oxidative phosphorylation was retained, and since sedimentation times were shorter with less concentrated sucrose, 0.25 M sucrose became the method of choice (Hogeboom, Schneider, and Palade, 1948). Mitochondria in this medium are spherical.

Biochemical and ultrastructural analyses soon showed that the fractions obtained were inhomogeneous. Harvey (1931) had suggested density gradient centrifugation might provide a more refined means of separating particles rather than by their size. This was introduced by Holter (1953) using sucrose solutions of varying density in a swinging bucket centrifuge. The crude cell fractions obtained by differential centrifugation were resuspended and layered over sucrose gradients. These could be set up to give continuously increasing density towards the bottom of the centrifuge tube, or, more commonly, layered to give several distinct bands. Equilibrium centrifugation now led to particles of similar size but with different density separating themselves appropriately. Recoveries in the first "differential" centrifugation were usually good (>80%); equilibrium sedimentation gave very clean fractions but poorer yields.

Some of the results obtained by differential centrifugation showed enzyme distribution between different cell fractions which were difficult to interpret. Enzymes like carbamoyl phosphate synthase or isocitrate dehydrogenase were found both in mitochondria and in the soluble fraction of the cell. This led to detailed kinetic studies with purified enzymes which indicated there might be populations of enzymes with slightly different properties (isozymes) catalyzing similar reactions in different compartments or in different cell types. Variations in kinetic behavior appeared to tailor the enzyme appropriately to the particular compartment or cell where the reaction took place.

Even greater uncertainty arose when only very small amounts of a constituent were detected outside the organelle in which it was principally concentrated. When the observation was reproducible, like the presence of small amounts of DNA (<1% of the total cellular DNA) in highly purified mitochondria, it required considerable work on the size, composition, and organization of mitochondrial DNA before its significance was accepted.

THE INTRACELLULAR ORGANELLES

Mitochondria

The importance of the availability of purified mitochondria for Lehninger's early studies of oxidative phosphorylation, and in the analysis of fat metabolism, has already been mentioned. Harman, in Green's laboratory, used a particulate preparation "cyclophorase" (1950) which catalyzed the reactions of the tricarboxylic acid cycle and was believed to contain the enzymes in some organized array. Biochemical properties and morphology confirmed cyclophorase and muscle sarcosomes, which had been isolated by Watanabe and Watson in 1951, were shown to be the equivalent of mitochondria in liver. By 1949 Claude described mitochondria as the "power plants of the cell."

Between 1953 and 1955 electron micrographs from Palade's and Porter's groups in the Rockefeller Institute and from Sjostrand's laboratory at the Karolinska Institute in Stockholm established the fine structure of the mitochondrion with its inner and outer membranes and its central matrix invaginated with folds from the inner membrane—the cristae. Initially there was some uncertainty over the origins of the infolds—whether or not the cristae were continuous from the inner membrane. In part, this arose from the difficulties of interpreting electron micrographs where the mitochondria had been sectioned obliquely. As more pictures became available it was easier to recognize the varying planes through which the mitochondria had been sectioned and so to analyze the micrographs more accurately.

Studies were also made of mitochondrial physiology. Laird noted that mitochondrial numbers were greater in metabolically active cells like liver (ca. 1000) compared with resting cells like small lymphocytes (<10). Keith Porter linked the extent of cristal surface with the amount of work done by the cell. Muscle mitochondria had significantly more cristae than those from liver.

As further tissues were examined it became evident that the details of mitochondrial morphology were very variable. While most cells had rod-or sausage-shaped organelles, some were spherical. Other cells had mitochondria with spiral cristae or with massive crystalline inclusions. In confirmation of earlier suggestions from classical microscopists the position of mitochondria in cells was also seen to be linked with the site in the cell where energy was required. In skeletal muscle the mitochondria were adjacent to the myofibrils; in the renal tubules they were close to the inner (non-luminal) surface of the cell which was then found to be the location of the Na/K-ATPase involved in active transport. The position of mitochondria in cells was not static however. Frederic (1951-1952) obtained striking phase contrast cinematographic films showing mitochondria in motile, dividing cells moving (or being moved) to the periphery of the nucleus immediately prior to cell division.

It soon became clear that the properties of the outer and inner membranes of the organelle were very different, with the outer membrane being apparently permeable and the inner membrane behaving as a classical semi-permeable membrane. This was confirmed as techniques were explored for "taking the mitochondria to pieces." The outer membrane could be selectively detached by very gentle sonication, extraction with low concentrations of detergent (deoxycholate) (Siekewitz and Watson, 1956), or even by using phospholipase very carefully to hydrolyze complex lipid in the outer membrane and so puncture it. With the outer membrane removed, demembranated mitochondria were sedimented and their composition and enzyme content determined. All the enzymes of the tricarboxylic acid cycle and for fatty acid oxidation were found in the defrocked mitochondria. When the inner membrane was ruptured, succinic dehydrogenase and the electron transport system (excluding cytochrome *c* which was easily dissociated—chapter 5) remained associated with the membranes. Other enzymes of the cycle were released and so assigned to the mitochondrial matrix. Some cytochemical methods could be adapted to the EM level, so that the presence of succinic dehydrogenase in the inner membrane was confirmed.

Lysosomes

Careful examination of the yellowish sediment obtained after spinning down the crude mitochondrial fraction showed it was frequently overlaid with loosely packed, fluffy material —the "fluffy layer." Experiments from de Duve's and, later, Novikoff's laboratories in the 1950s demonstrated that the lighter, lysosomal fraction was enriched in a number of hydrolases including acid phosphatase, aryl sulphatase, ß glucuronidase, RNAase, and a peptidase, cathepsin. All the enzymes had optimal pHs in the acid range (pH 5–pH 6). Density gradient centrifugation was then used to separate the heavier mitochondria from the lighter lysosomes.

If the original homogenization was very gentle hydrolase activity in the lysosomes was low, but was greatly enhanced if the fraction was treated with nonionic detergents such as Triton X-100. Lysosomes in many types of cells were recognized by Novikoff after acid phosphatase had been detected using the Gomori procedure. A role for these bags of hydrolases in intracellular digestion was suggested (De Duve) after lysosomes containing partially broken down mitochondria had been seen in the EM.

Some cells, like liver parenchymal cells, contained large numbers of particles in their cytoplasm. These ranged in size from mitochondria to glycogen granules, recognized by PAS staining. Extension of the careful analyses pioneered by de Duve's group showed that the cells contained another class of inclusions which were distinguishable cytochemically and could be isolated after lysosomes had been disrupted by detergents. The particles, microbodies, contained flavoprotein oxidases like D-amino acid oxidase (chapter 7) and catalase, which destroys hydrogen peroxide, the product of the oxidase activity.

Microsomes

The microsomal fraction was first obtained by Claude in 1943. In addition to lipid in the fraction, he noted the presence of RNA-rich granules, consistent with reports from Brachet that cytoplasm stained for RNA by the methyl-green/pyronin procedure. Glucose-6-phosphatase was a prominent enzyme when the fraction was prepared from liver. Since density gradient sedimentation showed G-6-P-ase was absent from mitochondria and lysosomes, it was used as a marker for liver microsomes.

Interpretation (1953-1955) of the ultrastructures which, after cell disruption and differential centrifugation, gave rise to microsomes, proved quite difficult. Porter described a vacuolar system with canaliculi and cisternae which was particularly evident in protein-secreting cells such as the acinar cells of the pancreas. The appearance was reminiscent of the ergastoplasm of classical microscopists (Garnier and Bouin, 1897). Palade first observed that regions between the cisternae were studded with small (10-15 nm) granules. He and Siekewitz (1956), Littlefield, Zamecknik and others used deoxycholate to remove lipid from the microsomal pellet and release the RNA/protein-containing granules—the ribosomes. Ribosomes were seen to be prominent in protein synthesising cells.

Early studies with labeled amino acids (Chapter 8) showed they were extremely rapidly incorporated into protein in the small granular fraction. Workers in Zamecknik's laboratory found this uptake was greatly diminished if the microsomes were treated with RNAase. It gradually emerged that there were two families of ribosomes: those which appeared to be attached to lipid membranes (the endoplasmic reticulum), and those which seemed to be free in the cytoplasm. Endoplasmic reticulum (ER) carrying ribosomes was designated rough ER. This was very evident in protein-secreting cells.

Electron microscopists, notably Sjostrand, were more concerned with the organization of the lipid-rich component of the microsomes. They noted the double lamellar leaflet structure could be very pronounced, as for example in the pancreas. Weiss, looking at these membranes ca.1953, described them as flattened sacks which might participate in the formation of zymogen granules. The function of the ER in post-translational processing of proteins to be secreted and the relation of the ER to the Golgi apparatus were to be intensively studied in the 1970s and 1980s.

The Golgi Apparatus

Golgi, in early neuroanatomical studies (1898) staining neurones by silver impregnation, observed a reticular apparatus which was crescent shaped and appeared to be linked through canaliculi. The structure was also seen in secretory cells. Between 1949 and 1954, Baker reported the presence of similar systems in unfixed cells examined by phase contrast. The structures could be stained by osmium tetroxide and probably contained lipid. They also stained for glycoprotein and alkaline phosphatase. Baker's confirmation of the existence of the

Golgi apparatus was greeted with considerable scepticism. However, with the introduction of the EM, the Golgi apparatus was convincingly demonstrated in snail spermatocytes, which showed duplex, rather closely packed, membrane structures with associated vesicles. Virtually the same regions were then seen in somatic cells.

The role and organization of the Golgi apparatus in processing proteins for transport to other intracellular organelles or for export (exocytosis) is still being investigated.

The Nucleus

Following the publication (1859) of Darwin's, *The Origin of Species,* it became necessary to identify the basis of evolution. With the rediscovery of Mendel's laws (1900) the idea arose that there were living units smaller than cells which might be below the limits of resolution of visible microscopy. These "micromeres" were postulated, *inter alia,* to carry the inheritable factors. Hertwig (1894) suggested the nucleus was the vector of inheritance. He used the term "ideoblast" to define "the smallest particles of material into which the hereditary mass ... can be divided ... Metaphorically they can be compared to letters of the alphabet. ... In my view the nucleus is the bearer of the ... hereditary material, [i.e.] of a substance which is more stable than protoplasm, and because it is less subject to influence [from] the outside world, it stamps its specific character on an organism."

Claude Bernard was the first to suggest that the cell nucleus was concerned with synthetic reactions, a view sustained by later experiments with amoebae. If amoebae were transected, only the nucleated halves could continue the syntheses essential for growth. Observations by Flemming (late 19th century) on the behavior of nuclei during fertilization and cell division are the basis of modern cell biology.

Nuclei were first isolated by Miescher (1869) from pus cells recovered from discarded surgical bandages. The principle constituent—a phosphorus-rich material then called nuclein—was stained by methyl green. A few years later salmon sperm were shown to contain a phosphorus-rich acidic compound—the nucleic acid—and a basic protein "protamine." Further work by Kossel, Levene,

Hammarsten, Gulland, and their colleagues led to the isolation from animal thymus glands of thymonucleic acid (DNA) containing a pentose sugar and the bases adenine, guanine, cytosine, and thymine. Yeast gave a nucleic acid with a different pentose (RNA), but the same bases except for the replacement of thymine by uracil. Wheat embryos were then shown to contain a nucleic acid resembling that in yeast so that by 1921 W. Jones could suggest "... there were only two nucleic acids in nature, one [only] obtainable from the nuclei of animal cells and the other [only] from the nuclei of plant cells." This simple view had to be abandoned in the 1930s when DNA was isolated from plant cells and yeast, and RNA was found in pancreas and other mammalian tissues.

In 1939, Behrens developed a nonaqueous technique for isolating cell nuclei. Reasoning that material could easily be lost from organelles if tissue was disrupted in aqueous medium, Behrens froze the samples in liquid nitrogen and ground them in organic solvents which precipitated protein on the surface of organelles thus preventing low molecular weight substances from escaping. The finely ground material was then partitioned in benzene/carbon tetrachloride gradients, with debris floating to the surface and the dense nuclei sedimenting. After a few rounds of partioning, purified nuclei were obtained and their enzymic activity investigated.

In 1948, Boivin, Vendrely, and Vendrely quantitatively determined the amount of DNA in bovine sperm and various other organs. The average amount of DNA/ diploid nucleus was 6-6.4 pg, approximately twice that found (3.3 pg) in the haploid sperm nuclei. Similar results were reported by Ris and Mirsky (1949) who showed adult liver to contain polyploid nuclei with 2n, 4n, and 8n amounts of DNA. Application of visible microspectrophotometry and Feulgen staining to the determination of DNA (Swift, 1950) led to the realization that, in a given species, nuclear DNA contents were identical even if nuclear volumes in different cells varied some 20-fold; i.e. nuclei contained varying amounts of protein. Microspectrophotometry also suggested DNA was synthesized detectably before the start of prophase, a result confirmed almost immediately by studies of isotope incorporation into DNA (Howard and Pelc, Chapter 8). Amounts of DNA were halved again after the chromosomes had separated to the poles of the spindle apparatus and cytokinesis had occurred. Once the cellular content of DNA was known, amounts of enzymes or other constituents could be expressed on a cell basis. It was realized that such a reference standard might be more appropriate than that on the much more commonly used protein or dry weight basis (Ennor and Stocken, 1947).

Isolation of nuclei by the nonaqueous technique is extremely tedious and many biochemical functions do not survive the exposure to organic solvents. First attempts to use aqueous solutions and the differential centrifugation technique of Hogeboom and Schneider were not encouraging. Not only were nuclei mixed up with only partially broken cells and erythrocytes but examination by phase contrast clearly showed nucleoprotein leaking out from nuclei. The leakage was very considerably reduced if 5 mM Ca^{2+} were present in the medium (Schneider et al., 1952). By 1956, Chauveau had developed a procedure in which the tissue, or the crude nuclear fraction, was sedimented through dense sucrose (2.2 M) /5 mM of calcium chloride. Clean colorless nuclei now sedimented while red cells, mitochondria, etc. remained at the top of the centrifuge tube. Experiments by Siebert and others showed DNA and histone contents of nuclei prepared in this way were not significantly different from those obtained by the Behrens method.

Another procedure used to isolate nuclei from some tissues was to disperse the material in dilute citric acid, (pH 6.2) from which clean nuclei could be sedimented (see Dounce and Umana, 1962). Certain enzymes, such as those synthesizing NAD^+, were significantly concentrated in the organelle, but generally leakage from nuclei was a major problem.

Ultrastructural examination of nuclei *in situ* showed they were not surrounded by a continuous double-layered membrane, but that the membrane was interrupted by pores (Callan and Tomlin, 1950). These were not holes but were highly organized structures involved in transport between the nucleus and the cytosol.

Attempts to use isolated nuclei for DNA or RNA synthesis were disappointing. Careful study of isotope uptake into RNA showed that, at best, ribonucleotides were only incorporated into transcripts which had already been initiated *in vivo*. It is only since the 1970s, after considerable work with microbial enzymes, that replication and transcription have been successfully analyzed in eukaryotes *in vitro*.

THE CELL (PLASMA) MEMBRANE

Most of the early work on membranes was based on experiments with erythrocytes. These cells were first described by Swammerdam in 1658 with a more detailed account being given by van Leeuwenhoek (1673). The existence of a cell (plasma) membrane with properties distinct from those of protoplasm followed from the work of Hamburger (1898) who showed that when placed in an isotonic solution of sodium chloride, erythrocytes behaved as osmometers with a semipermeable membrane. Hemolysis became a convenient indication of the penetration of solutes and water into the cell. From 1900 until the early 1960s studies on cell membranes fell into two main categories: increasingly sophisticated kinetic analyses of solute translocation, and rather less satisfactory examinations of membrane composition and organization.

Permeability Properties

Gryns (1896), Hedin (1897), and especially Overton (1900) looked at the permeability of a wide range of different compounds, particularly non-electrolytes, and showed that rates of penetration of solutes into erythrocytes increased with their lipid solubility. Overton correlated the rate of penetration of the solute with its partition coefficient between water and olive oil, which he took as a model for membrane composition. Some water-soluble molecules, particularly urea, entered erythrocytes faster than could be attributed to their lipid solubility—observations leading to the concept of "pores," or discontinuities in the membrane which allowed water-soluble molecules to penetrate. The need to postulate the existence of pores offered the first hint of a mosaic structure for the membrane. Jacobs (1932) and Huber and Orskov (1933) put results from the early permeability studies onto a quantitative basis and concluded molecular size was a factor in the rate of solute translocation.

From 1938 until the early 1960s Wilbrandt undertook extensive experiments on glucose uptake into erythrocytes. It was appreciated that the uptake by red blood cells (RBC) was atypical in respect of its insensitivity to insulin. Nevertheless the significantly greater rate at which the sugar could traverse the membrane than its maximum rate of utilization (Widdas, 1954) made studies with RBC easier to analyze than experiments on glucose uptake by yeast (e.g. Rothstein, 1956). Glucose uptake by erythrocytes was greater than expected from simple diffusion. When sugars of related structure were added to the medium (e.g. LeFevre, 1954) competition was evident and the translocation was stereochemically specific. It was therefore proposed that diffusion was somehow facilitated, with the glucose molecule crossing the membrane through water-filled channels or pores (Danielli, 1952). This interpretation could not however be distinguished kinetically from translocation aided by the presence in the membrane of carriers. Reduced translocation of solutes in the presence of reagents such as fluorodinitrobenzene or *p*-chlormercuribenzoate suggested proteins might be involved in the transport process.

Until 1939 it was believed that the plasma membrane was impermeable to cations (Moore, 1912; Jacobs, 1931), the asymmetric distribution of Na^+ and K^+ across the human erythrocyte membrane being a static property of the cell. Between 1939 and 1941 Hevesy and his associates used ^{24}Na and ^{42}K to show the membrane was permeable to both cations. Net movement of cations was reported by Maizels and Patterson (1940); human erythrocytes stored in the cold prior to transfusion gained Na^+ which was expelled *in vivo*. Sodium efflux and potassium influx were abolished by glycolytic inhibitors—iodoacetate (Maizels, 1951) or fluoride (Davson, 1941; Eckel, 1958). In the mid-1950s Glynn, Maizels, Post, and their colleagues established that active transport of cations occurred across the erythrocyte membrane, with a suggested stoichiometry of $3Na^+$ out to $2K^+$ in. Inhibition of cation fluxes by cardiac glycosides such as ouabain, was reported by Schatzmann in 1953.

Major developments in transport kinetics followed from the work of Gardos who, in 1954, succeeded in restoring K^+ uptake in red cell ghosts if ATP was added to the medium. Hoffman (1962) showed that in the presence of inosine, the ghosts extruded Na^+. Three components of efflux were distinguished: active transport, passive transport, and exchange diffusion. Active transport was dependent on ATP and was blocked by ouabain.

The experiments were continued by Hoffman and Whittam, who concluded that a protein, an ATPase, in the membrane was necessary for active transport and was vectorially organized, with ATP and Na^+ being required internally and K^+ externally where ouabain was inhibitory. The ATPase was finally identified as the "sodium pump" by Skou (1957); it vectorially translocated Na^+ and K^+ across the membrane, and was phosphorylated transiently in the process.

In the early 1960s Bangham prepared liposomes—sealed vesicles made from lecithin or other lipids, with an aqueous interior. Liposomes with purified Na/K-ATPase inserted into their membranes pumped cations in the presence of ATP, convincingly confirming the function of the ATPase, and the existence of vectorially (asymmetrically) organized proteins.

Membrane Structure

Until the 1960s the plasma membrane was virtually a "black box" across which selective translocation of solutes occurred. Kinetic studies provided descriptions of solute movements which were not dependent on the detailed structure of the membrane, but did offer parameters into which any model structure must be accommodated. Concomitantly with studies on kinetic properties, analyses were attempted of membrane constituents and their organization. Red cell ghosts resulting from lysis at neutral pH provided material for examination. The lytic procedure chosen determined the amount of hemoglobin remaining associated with the ghosts—usually 2-3% which gave a hemoglobin content of as much as 50% dry weight. Inevitably the membranes were randomly broken and collapsed onto one another. There was general agreement about gross lipid composition (30% phosphatidyl choline, 25% phosphatidyl ethanolamine, 15% phosphatidyl serine, 22% sphingomyelin, with traces of lysolecithin and phosphatidic acid). When gas chromatography became available in the late 1950s, oleic acid was found to be the most prominent unsaturated fatty acid (48.2% by weight). Of the total fatty acids, 22.6% were saturated. Isotope studies showed that both the lipid and protein components of mature, non-nucleated, mammalian erythrocyte membranes were stable. There was great difficulty in isolating and identifying the proteins.Methods using detergents were not yet in common use and attempts to diminish amounts of hemoglobin by washing with dilute citrate or acetate buffers caused protein denaturation and gave variable results.

As early as 1925 Gorter and Grendel had estimated from the amount of ether-soluble lipid in mammalian erythrocytes that there was just enough lipid which, if closely packed in a monomolecular film, would cover twice the surface area of the cells. They concluded there could be a lipid bileaflet membrane around the cell, a very influential postulate, though later shown to be based on erroneous calculations. Experiments of Langmuir and others confirmed that when lipids formed monolayers on water with molecules oriented perpendicular to the surface, their polar head groups entered the water phase. The structure of the membrane was below limits of resolution of visible microscopy. It was not until the 1950s that electron micrographs of cells stained with OsO_4 or $KMnO_4$ confirmed the bileaflet arrangement. An influential interpretation of a bileaflet membrane structure was that of Davson and Danielli (1935). Because the surface tension of cell membranes was less than would be expected with pure lipid, they postulated the presence of protein on either side of the lipid bilayer.

During the 1950s isotope studies coupled with differential centrifugation of cell fractions indicated that in membranes from nucleated (bird) erythrocytes, in contrast to those from mammalian RBC, lipids and proteins turned over rapidly and at different rates. In human erythrocytes it became apparent that the specificity of blood group substances, whose importance in transfusions was by then well understood, was a function of their glycosylation (see Watkins et al. 1981). No evidence for an ordered arrangement of proteins was forthcoming nor was the independent turnover of the membrane constituents easily accommodated into the Davson-Danielli picture. In 1959, therefore, Robertson advanced a new "unit" theory of membrane structure. He envisaged a lipid bileaflet with the charged head-groups of the complex lipids covered by glycosylated protein or mucopolysaccharide externally and extended polypeptide chains underlying the bilayer. Such a picture was consistent with the trilaminar appearance of the membrane in the EM and with Finean's low angle X-ray diffraction analyses of myelin sheaths. Electrophoretic studies established animal cells had negatively charged surfaces. Initially this was attributed to phosphate from phospholipids (Furchgott and Ponder, 1941; Bangham et al., 1958) but by the early 1960s the major source of the negative charge was shown to be the carboxyl group of sialic acid in surface glycoproteins.

By the late 1960s the unit theory in its turn became increasingly incompatible with observations on the diverse constituents, turnovers and transport characteristics of membranes in different cells. There was also no evidence that proteins were extended along the cytoplasmic face of the membrane. The introduction of freeze-fracture electron microscopy showed inclusions which penetrated lipid layers in the membrane (Marchesi et al., 1972). In erythrocytes, labeling studies (^{125}I-labeled lactoperoxidase and immunofluorescence) indicated the presence of integral proteins corresponding to these inclusions, which spanned the membrane and were asymmetrically accessible to proteolytic action. Other proteins, like cytochrome *c* on the outer face of the inner membrane of the mitochondrion, were peripheral and more loosely bound. Cell-fusion studies and lymphocyte capping in response to antigens demonstrated the membrane was not rigid. Components could move in the plane of the membrane. Consideration of these and other findings led Singer and Nicolson in 1972 to advance the fluid mosaic theory of membrane structure—the basis of current views.

The heyday of differential centrifugation and work on isolated cell fractions was short. In the 1940s and 1950s biochemists attracted much critical comment because of their use of disorganized homogenates and then of separated cell fractions. Many metabolic pathways were, however, successfully analyzed with little reference to their normal setting within the cell.

REFERENCES

Bangham, A.D. (1972). Lipid bilayers and biomembranes. Annu. Rev.Biochem. 41, 753-776.

Brachet, J. (1957). Biochemical Cytology. Academic Press, New York.

Davidson, J.N. (1950). The Biochemistry of the Nucleic Acids. Methuen, London.

Dounce, A.L. (1954). Significance of enzyme studies on isolated cell nuclei. Intern. Rev. Cytol. 3, 199-223.

Drury, R.A.B., Wallington, E.A., & Cameron, R. (1967). Carleton's Histological Techniques, 4th ed. Oxford University Press.

Duve, C. du & Berthet, J. (1954). Differential centrifugation. Intern. Rev. Cytol. 3, 225-275.

Finean, J.B. (1972). The development of ideas of membrane structure. Sub-Cell. Biochem. 1, 363-373.

Glick, D. (1949). Techniques of Histochemistry and Cytochemistry. Interscience, New York.

Glick, D. (1953). Approaches to quantitative histochemistry and cytochemistry. Intern. Rev. Cytol. 2, 447-474.

Hogeboom, G. (1951). Separation and properties of cell components. Fed. Proc. 10, 640-645.

Holter, H. (1952). Localization of enzymes in cytoplasm. Adv. Enzymol. 13, 1-20.

Mirsky, A.E. (1943). Chromosomes and nucleoproteins. Adv. Enzymol. 3, 1-34.

Pollister, A.W. & Pollister, P.F. (1957). The structure of the Golgi apparatus. Intern. Rev. Cytol. 6, 85-106.

Reuck, A.V.S. de & Cameron, M.P., Eds. (1963). Lysosomes. CIBA Foundation Symposium. J. & A. Churchill, London.

Singer, S.J. (1976). Molecular organization of membranes. Annu. Rev. Biochem. 43, 805-833.

Whittam, R. (1964). Transport and Diffusion in Red Blood Cells. Edward Arnold, London.

Chapter 10

CONCEPTS OF PROTEIN STRUCTURE AND FUNCTION

1800—1940

The word "protein" (Gk. "first rank or position") is thought to have been suggested by Berzelius to the Dutch scientist G.J.Mulder in a letter (July 10, 1838), to describe an organic substance, containing nitrogen, sulfur, and phosphorus, Mulder had earlier obtained from various biological sources. Berzelius wrote:

> Or je presume que l'oxyde organique, qui est la base de la fibrine et de l'albumine (et auquel il faut donner un nom particulier, par example, proteine) est compose d'un radical ternaire, combine avec de l'oxygene dans quelqu'un de ses rapports simples que la nature inorganique nous presente." . . . "il parait etre la substance primitive ou principale de la nutrition animale que les plantes preparent pour les herbivores et que ceux-ci fournissent ensuite aux carnassiers.

Much of the early work which would lead to the identification of proteins as defined chemical entities started from observations on enzymes, either those involved in fermentation or on the characterization of components in gastric secretions which powerfully catalyzed the hydrolysis of different foodstuffs. As well as the digestive enzymes, a number of relatively pure proteins could be isolated from natural sources where they made up the major component (Table 1). Because of the importance and difficulty of isolating pure proteins and demonstrating their homogeneity, functionally active and relatively abundant

Table 1. Early Sources of Purified Proteins

Protein	*Source*
Albumin	Serum
Casein	Milk
Collagen	Tendons
Fibroin	Silk
Gelatin	Tendons
Hemoglobin	Erythrocytes
Keratin	Hair, Wool
Myosin	Skeletal Muscle
Zein	Maize

soluble proteins like hemoglobin and the digestive enzymes proved most useful in establishing that proteins were defined macromolecules.

During the nineteenth century this was not believed. Graham (1861) had proposed that proteins were colloids, non-crystallizable, diffusing slowly and unable to penetrate certain membranes. Graham stated: "The colloid is in fact the dynamical state of matter, crystalloid being the static condition. The colloid possess 'energia.' It may be looked upon as the probable primary source of the force appearing in the phenomena of vitality." No distinction was made between covalently bonded molecules and "secondary valences" thought to cause smaller molecules to aggregate together.

Isolation of individual amino acids started about 1820; by 1904 all of the naturally occurring amino acids in proteins had been isolated except methionine (Mueller, 1922) and threonine (Rose, 1937). One of the earliest methods for the separation of amino acids was through the differential volatility of their methyl or ethyl esters (Emil Fischer, 1901). This approach led to the discovery of valine, proline, and hydroxyproline. [In the 1970s Fischer's method was modified for microanalysis of proteins, separating the amino acid esters by gas phase chromatography. Separation is now usually performed by hplc (high pressure liquid chromatography).]

Some individual amino acids could be detected by specific color reactions such as the Millon reaction for tyrosine (1849), the Hopkins-

Cole test (1901) for tryptophan, and the Sakaguchi reaction for arginine (1925).

The method of protein hydrolysis was important; acid hydrolysis caused destruction of tryptophan but alkaline treatment gave even greater losses of other amino acids especially cystine. The amino acids were usually separated by then standard chemical procedures based on differences in solubility, selective precipitation by agents such as Reinecke salt (proline and hydroxyproline), or flavianic acid (arginine).

Quantitative recovery of the amino acids was virtually impossible with the classical methods. The total N-content of protein, (ca. 16% by weight) could be determined by micro-Kjeldahl analysis. Without accurate recoveries it was impossible even to determine amino acid contents of proteins. Heroic efforts were made by Chibnall and his collaborators at Imperial College in London to determine amounts of amino acids in pure proteins. By the early 1940s they had succeeded in accounting for 93.4% by weight of the amino acids in insulin. Other proteins gave lower figures. Complete analysis was seriously handicapped by the losses inherent in the hydrolysis techniques and the inadequate separation methods. Growth assays using microorganisms with specific dependencies on individual amino acids were attempted (e.g. *L.casei* for arginine), but were time-consuming and not very precise. The first complete analysis of the amino acid content of a protein, ß-lactoglobulin, was achieved by Brand in 1945.

Chibnall succeeded F.G. Hopkins in the Chair of Biochemistry in Cambridge in 1943. His appreciation of the need for accurate amino acid analysis underpinned the achievements there of Sanger and his pupils in determining the first amino acid sequence in a protein (see below).

Many theories were proposed of how the amino acids were associated in the protein structure. When corrections had been made for the ε-amino group of lysine and the dicarboxylic acid residues from glutamic and aspartic acids, recoveries of amino acids after hydrolysis were sufficiently good to indicate the equivalence of α-amino and carboxylic acid groups, thus suggesting that branching of the peptide chains was unlikely. By 1905, Emil Fischer and Hofmeister independently postulated that amino acids were joined together by peptide bonds to form long chains.

Fischer then synthesized a number of di-and tripeptides and showed that their properties were identical to those of the di-and tripeptides which could be obtained after partial protein hydrolysis. A peptide containing 18 amino acid residues was eventually synthesized. Molecules containing peptide bonds were found to give a characteristic pink to purple color in the presence of dilute alkaline copper sulfate. The simplest compound which does this is biuret, formed when urea is heated at 150-160 °C:

$$2CO(NH_2)_2 \text{ ----> } H_2N{\cdot}CO{\cdot}NH{\cdot}CO{\cdot}NH_2 + NH_3$$

The biuret test became a standard method by which to detect and estimate protein.

Early Methods of Protein Isolation

Progress in the determination of protein structure presupposed the availability of appropriate techniques for the isolation from natural sources of proteins which retained their normal functions (remained "native") and for the separation and assay of the component of interest. Almost all the early procedures involved "salting-out," the selective precipitation of proteins from aqueous solution by the addition of increasing amounts of salts. Later the converse process, "salting-in," was also employed. Because of its solubility in water, ammonium sulfate was commonly used although its solubility is quite sensitive to temperature changes between 0 and 20 °C. It is also a poor buffer so that separations were highly sensitive to changes in pH. Further complications ensued if the protein was to be estimated by the Kjeldahl process where nitrogen was converted to ammonium sulfate.

Precipitation by ethanol in the cold was used effectively by J.Mellanby in 1908 to obtain diptheria antitoxin; two years later Hardy and Gardiner reported the precipitation of plasma proteins by cold ethanol or acetone. The resulting proteins remained soluble in water, i.e. they were not denatured, and subsequent estimation of protein as nitrogen was helped by the use of nitrogen-free precipitants. Separations using organic solvents were considerably extended by Edsall,

Cohn, and their associates in Harvard during World War II in their efforts to isolate serum proteins for clinical use. Traces of the organic solvents were removed during the freeze-drying of the different preparations. In the U.K. plasma fibrinogen was obtained by Kekwick and his colleagues by low-temperature precipitation with ether. Serum proteins acquired by these chemical methods were mixtures.

Greater purification could be achieved, usually on a smaller scale, if the desired protein had easily measurable properties (e.g. was an enzyme), so that the specific activity of the product and extent of purification could be estimated. Well into the 1950s any enzyme required in a laboratory had first to be isolated by those needing it. With increased knowledge of enzyme properties two further methods of purification were commonly tried: heat denaturation of contaminating proteins (with the incidental discovery of some remarkably heat-resistant enzymes), and protein precipitation at the iso-electric point.

pH

Sorensen is usually considered to be the first to have realized the importance of hydrogen ion concentration in cells and in the solutions in which the properties of cell components were to be studied. He is also credited with the introduction of the pH scale. Electrochemistry started at the end of the nineteenth century. By 1909, Sorensen had introduced a series of dyes whose color changes were related to the pH of the solution, which was determined by the H^+ electrode. The dyes were salts of weak acids or weak bases. He also devised simple methods for preparing phosphate buffer solutions covering the pH range 6-8. Eventually buffers and indicators were provided covering virtually the whole pH range.

By 1930 the H^+ electrode, which was not suitable for biological situations when CO_2/HCO_3^- were present, was replaced by the glass electrode (Hughes, MacInnes, and Dole). This came into routine use in biochemical laboratories in the 1940s, giving an accuracy of 0.01 pH unit compared with 0.1 unit obtained colorimetrically.

The concept of pH and the ease with which it could be measured allowed the electrochemical ideas of Debye and Huckel to be applied to polyvalent acids and ampholytes. *On the Ionization of Proteins* was published by Lindstrom-Lang in 1924, showing that proteins were amphoteric and could be titrated over a range of pH and their iso-electric points determined.

Protein Homogeneity

Determination of the structure of the protein required that the material examined was pure. The first relatively pure proteins to be obtained were hemoglobin and ovalbumin. The only test of purity available before the introduction of the ultracentrifuge or electrophoresis was that developed by Sorensen (1917). Using ovalbumin he showed that at a given temperature, pressure, and pH a pure protein obeyed Gibbs' phase rule, behaving as a single component with a precise solubility in a variety of salt solutions. A pure protein in solution in dilute ammonium sulphate showed a sharp decrease in solubility (salted-out) as the concentration of the salt increased. The concentration of ammonium sulfate at which this occurred was a characteristic of a given protein, i.e. the protein showed behavior predicted for a defined molecular species.

Some of the purified proteins had apparently simple amino acid compositions— for example, fibroin from silk containing 50% glycine, 25% alanine, and about 6% tyrosine; basic amino acids could also be detected. In the small basic proteins of fish sperm—salmine from salmon and clupeine from herrings—Kossel (1904) found 85-90% of the nitrogen was derived from arginine. Enzymic hydrolysis yielded tripeptides of the general formula X-Arg-Arg, suggesting the peptide chain might contain significant runs of -X-Arg-Arg-X-Arg-Arg.

To get from the amino acid analysis of the protein to its structure, its molecular weight had to be estimated. As early as 1885 Zinoffsky had reported a minimum molecular weight for hemoglobin of 16.73 kDa based on its elementary composition and assuming one atom of iron/mole, a remarkably accurate figure. For most proteins because of their large size and ease of denaturation, classical cryoscopic or ebulliscopic methods were impracticable. Osmotic pressure measurements were therefore used (Reid,1904, 1905) where differences in pressure were recorded between the protein solution contained within a semi-permeable membrane, and its dialysate. Great care was needed to ensure equilibrium had been obtained without denaturation or breakdown of the protein. Adair (1925) obtained a value of 67 kDa for hemoglobin, very close to that found with modern methods. Osmotic pressure determinations were not accurate with molecular weights greater than about 100 kDa.

Ultracentrifugal techniques (up to 10^5 g) to measure molecular weights were introduced by Svedberg from 1923.Equilibrium sedimentation was used to determine the molecular weights of horse carboxyhemoglobin and methemoglobin. In 1926, Svedberg and Fahraeus reported a value of 66.8 kDa, virtually identical to that found by Adair. With sedimentation velocity methods the number and sharpness of the boundaries as the protein moved down the tube demonstrated the degree of homogeneity of the preparation. Ovalbumin and hemoglobin were shown to be monodisperse consistent with accumulating evidence that proteins were defined macromolecules.

Early ultracentrifuges were extremely expensive and were therefore located in only a small number of laboratories. Nevertheless, rapid strides were made in the theory and practice of ultracentrifugation. By 1940 the behavior of proteins with molecular weights ranging from that of cytochrome *c* (13.4 kDa) to serum globulin (ca.170 kDa) had been studied.

The third test of protein homogeneity, developments from which remain in common use, was that of electrophoresis. Arne Tiselius had been a research assistant in Svedberg's laboratory. From 1925 he pioneered the application of electrophoresis to the analysis and separation of protein mixtures, showing with dialyzed serum differences in mobility of the protein components and the presence of three classes of globulins, α, ß, and γ.

Electrophoretic separations in the purely liquid phase, as used by Tiselius, were complicated by the need to keep stable boundary conditions. Problems arose from convectional and gravitational mixing. In the 1950s various attempts were made to increase the stability and thus the reproducibility of the procedure by using filter paper, or more effectively, cellulose acetate strips as support for the aqueous phase. Considerable success was achieved by Oliver Smithies (1955-1960) who introduced potato starch as the support medium. Acrylamide gels were developed from 1959 (Raymond and Weintraub; Ornstein, 1964; Maizels, 1966) and proved simple to use, highly reproducible and applicable to the separation of very small (μg) amounts of protein.

The most rigorous evidence that proteins had defined structures was probably the molecular weight determinations of Adair and Svedberg. From 1900, however the crystallization of increasing numbers of proteins, while not a very reliable indication of purity, suggested to Schulz that proteins were not colloidal aggregates but large molecules with definite structures.

This recognition of proteins as "macromolecules" was staunchly advocated by Staudinger (1920 et seq.) in spite of vociferous opposition: "Organic molecules with more than forty C atoms do not exist." "Purify your products . . . they will prove to be lower molecular weight substances." Even as late as 1938, Gorter maintained "All of the reactions and interactions which we call life take place in colloidal systems."

The ability to determine molecular weights prompted speculation about the structure of proteins. Abderhalden (1924) thought amino acids condensed to form substituted diketopiperazine rings. In 1936, Dorothy Wrinch proposed the "cyclol" hypothesis. This assumed polypeptide chains folded into hexagonal rings. The idea had to be abandoned when it was realized the folds would not accommodate the amino acid sidechains. From his early estimates of molecular weights Svedberg suggested proteins might be built up from units of 34 kDa; by 1939 this had been modified to $[17.6]_n$.

Analysis of protein shapes by X-ray crystallography was pioneered by Astbury in the 1930s. Fibrous proteins like fibroin, keratin, and myosin were examined. Keratin showed two patterns of diffraction, the α form, and a stretched form designated ß-keratin, which gave a pattern similar to that seen with fibroin. Myosin could be drawn into threads with X-ray pictures similar to those seen with intact muscle, fibroin, and ß keratin. Collagen, which is rigid and inextensible, gave a different pattern, later shown to originate from a triple helix (Crick and Rich, 1955).

Globular proteins were much more difficult to prepare in an ordered form. In 1934, Bernal and Crowfoot (Hodgkin) found that crystals were better preserved if they were kept in contact with their mother liquor sealed in thin-walled glass capillaries. By the early 1940s crystal classes and unit cell dimensions had been determined for insulin, horse haemoglobin, RNAase, pepsin, and chymotrypsin. Complete resolution of the structures required identification of the crystal axes and some knowledge of the amino acid sequence of the protein—requirements which could not be met until the 1950s.

In 1939 Corey began investigating the crystal structure of polyglycine and other simple polypeptides. From these studies and from model building, he and Pauling in 1951 suggested amino acids in proteins might be ordered as an α-helix or ß-pleated sheet. When the shapes of myoglobin and hemoglobin were resolved a few years later (see below) the existence of such structures was confirmed.

In spite of the inability to obtain accurate amino acid analyses, by 1940 it was generally recognized that proteins were high molecular weight charged molecules containing long chains of amino acids linked by peptide bonds. In a number of cases the protein molecules associated in regular arrays which could potentially be examined crystallographically. It was also realized (Edsall and Cohn, 1943) that specificity in the physiological behavior of individual proteins depended in large part on the arrangement in space of the amino acid side chains.

THE INTRODUCTION OF CHROMATOGRAPHY: THE ANALYTICAL REVOLUTION

The primary structure of the protein—the arrangement of the individual amino acids along the peptide chain—required the development of accurate, sensitive, and specific methods for recognizing and measuring the different constituents. The ends of the protein chain had to be identified and alternative procedures for hydrolyzing the protein so that overlapping peptides resulted and could be aligned. All these followed in the 12 years between 1941 and 1953, as techniques in chromatography were introduced and applied to the field of protein analysis.

That solutes could be separated by their different rates of migration up paper dipped in a solution of the compounds was first observed by Schoenbein and applied by Goppelsroeder (1861-1862) to the effective separation "of mixtures of dozens of compounds" (Jossang, 1992). The best known early application of chromatography is that of Tsvett, a physical chemist who was trained in Switzerland and went to Russia to work on plant physiology. In 1910 he published, *Chromophylls in the Plant and Animal Worlds,* which included a discussion on theoretical aspects of adsorption and reported the separation of chlorophylls α and β following their extraction by organic solvents. Tsvett reviewed liquid/liquid partition chromatography, but not its extension to countercurrent methods. He also contemplated the possible separation of solutes on filter paper but considered his methods using columns of adsorbent superior. Tsvett's studies were halted by World War I; he died in 1919.

Quantitative Analysis of Amino Acids

Great difficulties were experienced in laboratories trying to extend Tsvett's approach. Activated charcoal, Cγ alumina, and silica gels provided a range of adsorbents with different properties. Irreproducibility between different batches of adsorbent were almost inevitable, and very irritating, before the theoretical complexities of adsorption analysis were understood. Operationally, there were serious difficulties in the nondestructive monitoring of the elution of colorless solutes like amino acids.

Liquid/liquid partition chromatography was explored by Willstatter from 1913. The process was extensively developed by Martin and Synge (ca. 1941-1948) who partitioned amino acid derivatives between chloroform and water using precipitated silica as support for the aqueous phase. The preparations of silica were again very variable and it was difficult to prevent adsorption which interfered with the expected behavior of the aminoacids. At first methyl orange was added to the water phase to visualize the amino acids; the separation of the acids then caused a red band to move down the columns. The quantitative reaction with ninhydrin was introduced by Moore and Stein in 1948 for both the detection and estimation of the amino acids. Considerable attention was given to theoretical aspects of partition chromatography and the conditions under which the solutes behaved ideally. The R_F emerged as a quantitative characteristic of the behavior of a particular solute in a given solvent system. Such information allowed predictions about the feasibility of separating related substances so that Craig in the late 1940s and early 1950s, using countercurrent liquid chromatography, could purify a range of low molecular weight peptides and other substances. This method had only limited application in the protein field since both the organic solvents used and the continuous exposure of the solutes to the solvent interface favored denaturation.

The introduction of filter paper rather than silica to provide a solid, apparently inert support for the partition of solutes between water held in the pores of the paper as stationary phase, and the mobile organic solvent, prompted an analytical explosion (Gordon, Martin, and Synge, 1943; Consden, Gordon, and Martin, 1944). Reproducibility was ensured because virtually all workers used treated Whatman filter paper. [Later analysis by Martin (1950) suggested the paper formed a complex gel of hydrated polysaccharide with adsorptive properties.] While the method was originally reported for amino acids it was rapidly applied to the separation of many water-soluble low molecular weight compounds, provided they could subsequently be detected by suitable colorimetric reactions or by absorption in UV. Detection by UV was increasingly used as paper chromatography was applied to the separation of nucleic acid derivatives (Markham and Smith, 1949).

Paper chromatography had some obvious drawbacks. It was excellent for rapid qualitative analysis of components in a protein hydrolysate. Amino acids were visualized by reaction with various reagents but their quantitative assay was much more difficult. Estimation of the relative amounts of the constituents by densitometer scanning was unsophisticated and unsatisfactory. Reaction with ninhydrin was destructive, and although elution could be used on parallel chromatograms, it was tedious and not very accurate. Paper chromatography was very successful in separating small (μg) amounts of material. Thicker (3 mm) papers were introduced to allow larger quantities to be analyzed. For many purposes, however, column separations were more desirable because the components were serially eluted and the size of the columns could be adjusted to the weight of starting material.

Developments in the uses of synthetic ion exchangers had become available following World War II. The commercial materials were much more reproducible than those used earlier, being defined in particle size, cross-linking, and amount and type of ion exchanger present. Waldo Cohn (1949) separated nucleic acid derivatives on sulfonated polystyrene resins (Dowex-50), or on resins carrying substituted ammonium salts (Dowexs 1 and 2). In the U.K., Amberlite resins IR-100 and IR-4 were used. Cellulose-based ion exchangers were also introduced. In parallel with these developments in column chromatography, automatic fraction collectors came onto the market. Automatic monitoring in UV came later.

From 1951, Moore and Stein at the Rockefeller Institute refined the quantitative separation of amino acids on Dowex-50 which led to fully automated amino acid analyses. In early models two columns were needed: one of 100 cm to separate most of the acidic and monobasic monocarboxylic acids between pH 3-11; and a short, 15 cm column for the basic amino acids which were eluted at pH <7. The columns operated above room temperature to give more rapid results, and the elution was monitored automatically by quantitative ninhydrin reactions. By the late 1950s a protein hydrolysate could be analyzed overnight.

The Primary Sequence of Insulin

With methods for the quantitative analysis of amino acids to hand, the way was now open for the determination of amino acid sequences. Purified bovine insulin was relatively freely available. On the basis of ultracentrifugal analysis (Gutfreund and Ogston), a molecular weight of 12,000 was assumed—as it later emerged, a factor of 2 too high. One of the advantages from the choice of insulin as the protein to sequence was that tryptophan is absent. A 100% recovery of the amino acids could therefore be obtained easily by simple hydrolysis with HCl. In 1948 Tristram reported the complete amino acid composition of the protein.

Rapid deamination of α-NH_2 groups by nitrous acid was used by Van Slyke to determine the number of peptide chains present in proteins. Methods for the precise identification of end-groups were developed by Jensen and Evans (1935); treatment with phenylisocyanate indicated one of the N-terminal amino acids in insulin was phenylalanine. This was confirmed and extended by Sanger (Sanger, 1945; Porter and Sanger, 1948) who used fluorodinitrobenzene (FDNB) to label the reactive N-terminal α-NH_2 group, yielding a yellow dinitrophenyl derivative which was originally separated on silica gel. Glycine as well as phenylalanine were found to be N-terminal, so that for a molecular mass of 12 kDa there were four peptide chains, two A chains with N-terminal glycine and two B with phenylalanine. Sanger further established that ε-NH_2 groups of lysine reacted with FDNB more slowly than the α-NH_2 groups. The total lysine content in insulin and that determined from the ε-NH_2 groups were identical, i.e. there was no branching through the ε-NH_2 group.

In 1950 an alternative to the Sanger procedure for identifying N-terminal amino acids was reported by Edman—reaction with phenylisothiocyanate to give a phenylthiocarbamide labeled peptide. When this was heated in anhydrous HCl in nitromethane, phenylthiohydantoin was split off, releasing the free α-NH_2 group of the amino acid in position 2 in the sequence. While initially the FDNB method was probably the more popular, the quantitative precision which could be obtained by the Edman degradation has been successfully adapted to the automatic analysis of peptides in sequenators.

Identification of carboxy-terminal amino acids was also attempted. Studies by Bergmann and his associates in the 1930s (see below) had characterized various peptidases with differing specificities. One of these was carboxypeptidase which required a free carboxy terminus adjacent to the peptide bond to be hydrolyzed. The specificity of the enzyme was limited but Lens in 1949 reported alanine to be at one end of insulin. Fromageot and his colleagues (1950) and Chibnall and Rees (1951) reduced the carboxy termini to ß-aminoalcohols and showed glycine as well as alanine to be carboxy-terminal. Hydrazinolysis was also attempted; the dry protein was treated with hydrazine at 100 °C for 6 h so that the carboxy-terminal amino acid was released as the free acid. Cysteine and tryptophan, however, were degraded by this procedure and indeed none of the methods for determining the carboxy-terminal amino acid was universally applicable.

The A and B peptide chains in insulin are linked through disulfide bridges. Their presence was suspected from the change in molecular weight which followed the reduction of insulin. For quantitative analyses the S-S bridges had to be broken. Sanger, following the approach used by Toennies and Homiller (1942), oxidized the protein with performic acid, so that the half-cystines were converted to cysteic acid. After oxidation, insulin could be separated into its A and B chains, the A peptide with 20 amino acid residues and the B with 30.

For some purposes an alternative method, reducing the S-S link, was used (Moore et al.,1954). The protein was dissolved in 8 M urea, reduced with borohydride and the free thiol groups then treated with iodoacetate. Carboxymethyl cysteine could then be estimated chromatographically.

Next the A and B peptides were each attacked by partial acid hydrolysis and the relatively small peptides which resulted, separated by two-dimensional paper chromatography and visualized either by ninhydrin or by their fluorescence after heating the paper. The individual peptides were then eluted and their N-terminal and total amino acid composition determined. To obtain larger peptides with up to 15 residues, trypsin (Sanger and Tuppy,1951) and later chymotrypsin (Sanger, 1952) were used. These enzymes were available pure and crystalline, and although in Bergmann's laboratory there was considerable interest in the possibility that peptides were synthesized by the reversal of enzyme-catalyzed proteolysis (Fruton, 1950),there was no evidence for transpeptidation resulting from the action of hydrolases on insulin.

Trypsin specifically hydrolyzes peptide bonds with basic amino acid residues adjacent to the bond hydrolyzed. A family of peptides was released by trypsin hydrolysis, all except the original carboxy-terminal peptide having arginine or lysine in the carboxy position. The peptides were separated on paper by high-voltage electrophoresis followed by elution in the second dimension with organic solvents. Later, treatment of the protein or peptide with agents blocking the ε-NH_2 group of lysine allowed only bonds adjacent to arginine to be split. Chymotrypsin preferentially hydrolyzes peptide bonds downstream from aromatic residues, yielding a different, but potentially overlapping family of peptides from trypsin.

By 1953 the complete primary sequence of insulin was known. Sanger could thus conclude that proteins like insulin had unique structures and were, as Staudinger had postulated, defined macromolecules. The primary sequence of other proteins soon followed.

THE THREE-DIMENSIONAL STRUCTURE OF INSULIN

In spite of the early studies showing proteins could be crystallized and the recognition of the repeating units in, for example, insulin, there were major difficulties—technical, theoretical, and computational—in interpreting the X-ray diffraction patterns (see Crick and Kendrew, 1957). From the X-ray photographs the dimensions of the unit cell could be calculated but not the phase angle of the crystal, without which unequivocal interpretation of the structure was impossible. Two very different approaches were therefore explored. The first, which became especially applicable if some or all of the primary sequence was known, was to compute the expected diffraction pattern from a postulated three-dimensional structure, and to compare this to the pattern actually obtained from the crystals.

Alternatively or additionally, atoms were introduced into the molecule whose X-ray diffraction pattern could be recognized—the technique of isomorphous replacement (Green, Ingram, and Perutz, 1954). Cysteine SH groups were most commonly modified, conveniently by mercury, which is heavy enough to provide a reference point. For horse hemoglobin whose structure was to be solved by Perutz and his colleagues in the 1960s, six isomorphous heavy atom derivatives were made, giving 140 crystals and 40,000 reflections for analysis. Whale myoglobin, with only one heme group and approximately one-quarter the size of hemoglobin, had no free cysteines but gave a number of suitable crystalline derivatives (Bluhm, Bodo, Dintzis, and Kendrew, 1958). By 1961 its complete atomic structure was reported by Kendrew and his associates.

The three-dimensional structure of insulin remained recalcitrant in spite of the knowledge of its primary sequence. The early crystals had been found by Scott (1936) to contain zinc which could be replaced by other divalent metals. The zinc atom is not heavy enough to be unambiguously distinguishable. Eventually it proved possible to introduce uranyl acetate and uranyl fluoride into the insulin molecule and to obtain the three-dimensional structure, first at 2.8 Å resolution and then at 1.9 Å (see Blundell, Dodson, Hodgkin, and Mercola, 1972).

ENZYMES

Work on fermentation and the identification of enzymes in digestion has already been mentioned. In 1783, Spallanzani, following earlier work by Reamur (1752), had confirmed that meat fed to hawks and other fauna was liquefied by their gastric juice, a reaction which could continue outside the stomach. Payen and Persoz (1833) may have been the first to recognize the existence of an enzyme. Alcohol precipitated an agent from an aqueous extract of malt which converted malt to sugar. The agent was inactivated by heat and was called "diastase" (Gk. "separation") because it separated soluble sugar from the insoluble covering over starch granules. A similar activity was detected in saliva (amylase) and in 1836 Schwann extracted pepsin from the stomach wall.

Catalytic activities in plant extracts were also recognized early in the nineteenth century. Robiquet and Boutron (1830), for example, noted the hydrolysis of amygdalin, a glycoside in the seeds of bitter almonds; Robiquet, and Liebig and Wohler then identified the enzyme responsible—emulsin.

Enzyme Kinetics

Using fermentation and other examples from plant and animal sources, Berzelius in 1837 introduced the concept of catalysis, stating:

> This is a new force, belonging to both inorganic and organic Nature, for evoking chemical activity;...the nature of which is still concealed from us...I will call it

> the "catalytic power" of bodies, and decomposition by its means I will call "catalysis." In living plants and animals thousands of catalytic processes take place, giving rise to multitudes of chemical compounds. The cause of the phenomena in [the] living animal lies so deeply hidden that it certainly will never be found.

The controversy between the views of Liebig, Schwann and Pasteur over the nature of "ferments" was reviewed in Chapter 2. In part, this arose from confusion between "organized ferments," the living cells themselves, as exemplified by yeast, and unorganized ferments such as emulsin or pepsin which could work independently. The term, "enzyme," was coined by Kuhne in 1878 to describe the agent... "which occurs [in yeast] to exert this or that activity which is considered to be... fermentation." The study of enzymes has formed an important bridgehead for physicochemical exploration of biological systems from the time of Berzelius.

Quantitative measurements of simple and enzyme-catalyzed reaction rates were under way by the 1850s. In that year Wilhelmy derived first order equations for acid-catalyzed hydrolysis of sucrose which he could follow by the inversion of rotation of plane polarized light. Berthellot (1862) derived second-order equations for the rates of ester formation and, shortly after, Harcourt observed that rates of reaction doubled for each 10 °C rise in temperature. Guldberg and Waage (1864-67) demonstrated that the equilibrium of the reaction was affected by the concentration(s) of the reacting substance(s). By 1877 Arrhenius had derived the definition of the equilbrium constant for a reaction from the rate constants of the forward and backward reactions. Ostwald in 1884 showed that sucrose and ester hydrolyses were affected by H^+ concentration (pH).

Applications of chemical kinetics to enzyme-catalyzed reactions soon followed. Because of the ease with which its progress could be monitored polarimetrically, enzyme hydrolysis of sucrose by invertase was a popular system for study. O'Sullivan and Tompson (1890) concluded that the reaction obeyed the "Law of Mass Action" and in a paper entitled, "Invertase: A Contribution to the History of an Enzyme or Unorganized Ferment", they wrote "[Enzymes] possess a life function without life. Is there anything [in their actions] which can be distinguished from ordinary chemical action?"

Further experiments by Brown and particularly Henri were made with invertase. At that time the pH of the reactions was not controlled, mutarotation did not proceed to completion, and it is no longer possible to identify how much enzyme was used (Segal, 1959). Nevertheless, in a critical review of kinetic studies with invertase, Henri concluded (1903) that the rate of reaction was proportional to the amount of enzyme. He also stated that the equilibrium of the enzyme-catalyzed reaction was unaffected by the presence of the catalyst, whose concentration remained unchanged even after 10 hours of activity. When the concentration of the substrate [S] was sufficiently high the velocity became independent of [S]. Henri derived an equation relating the observed initial velocity of the reaction, V_0, to the initial concentration of the substrate, $[S_0]$, the equilibrium constant for the formation of an enzyme-substrate complex, K_s, and the rate of formation of the products, k_3:

$$V_0 = k_3 \cdot [S_0] / \, (1 + [S_0]/K_s)$$

In 1913 invertase was again employed by Michaelis and Menten, this time in acetate buffer. Initial velocities were measured and an expression for the Michaelis constant and V_{max} derived. $[S]$ was assumed to be $>>>[E]$, so that:

$$V_0 = k_3 \cdot [E_0] \cdot [S_0] / \, (K_s + [S_0])$$

A similar relation was obtained independently by Van Slyke and Zacharias (1914). They also considered enzyme inhibition and distinguished competitive inhibitors where the formation of an ES complex was "retarded," from instances where inhibitors interfered with the breakdown of the ES complex to give the products.

Henri and Michaelis-Menten kinetics assumed that the rate of formation of products was much less than that for the back reaction from ES to yield E + S. Van Slyke assumed the reverse. A more rigorous formulation was offered by Briggs and Haldane (1925) using steady-state assumptions previously applied to chemical kinetics by Bodenstein (1913).

The concept of ordered interactions of substrates with the enzyme and ordered dissociation of the products was advanced by Koshland in 1954. From then through the 1960s the introduction of stopped-flow techniques and relaxation methods allowed rapid reactions to be followed and the identification of transient intermediates, from which much more complex kinetic analyses have emerged (Fersht,1977).

The Nature of Enzyme Catalysis: Enzyme Specificity

In his consideration of the nature of catalysis Berzelius had assumed the catalyst played no part in the actual reaction. Studies on non-enzyme catalysis, and especially the roles of finely divided metals, such as platinum, seemed to substantiate this—a view apparently consistent with the concept of the adsorption isotherm introduced by Langmuir (1916).

Studies on the structure and mutarotation of mono- and disaccharides and on the specificity of their enzymic hydrolysis by glycosidases led Emil Fischer to propose the "lock and key" analogy for the interaction of an enzyme with its substrate (1894). The nature of the enyzme molecule itself remained controversial and was further confused by the uncertainty over the structure of proteins. Stopes in 1885 wrote "The reality of diastase is one thing, the substantial existence of a definite compound entitled to that distinctive name, is another." Bunsen and Hufner believed enzymes were proteins which were able to make temporary combinations with fermentable substances, and could be regenerated by splitting these combinations. Nageli (1879), de Jager (1890), and Arthus (1896) on the other hand thought enzymes were not definite molecules but were properties which could be associated with material substances, a view easily accommodated with Graham's views on proteins as colloids.

Standard tests, dialysis, heat inactivation, and the effects of changing the pH on the catalytic activity of the preparations, were all consistent with the idea that enzymes had the properties ascribed to proteins. Between 1920 and 1930 increasing numbers of enzymes were isolated by Willstatter and partially purified. Assays for purification and the extent achieved were not sufficiently rigorous to exclude the possibility that the active material was adsorbed on the surface of the protein (or colloid) and that the catalyst itself was not a protein. Even after Sumner in 1926 crystallized urease, the problem was not completely resolved because the low concentrations of the moderately pure enzyme which were catalytically active, were too dilute to give unequivocal positive tests for protein (not very sensitive at that time). In 1930 in his influential monograph, *Enzymes,* J.B.S. Haldane concluded "Enzymes [are] soluble, colloidal, organic catalysts produced by living organisms."

It was not until the 1930s when Northrop and his colleagues at the Rockefeller Institute crystallized a number of digestive enzymes (Northrop, Kunitz and Herriott, 1939; *Crystalline Enzymes*) that it was accepted that the catalytic activity of the enzyme was a property of the protein itself. The availability of these purified proteolytic enzymes enabled Bergmann and his associates in the 1930s to make detailed studies of the substrate parameters defining the specificities within this group of hydrolases. A range of synthetic peptides was tested as substrates, *exo*- and *endo*-peptidases were distinguished by their dependence or otherwise on the presence of free α-NH_2 or carboxy-terminal groups adjacent to the susceptible bond. A further consequence of the work was the demonstration that "proteinases" (enzymes hydrolyzing high molecular weight protein substrates), which included pepsin, trypsin, and papain, could equally easily utilize di- or tripeptides. Molecular weight was not a determinant in their specificity (Bergmann and Fruton, 1941).

Analyses of enzyme reaction rates continued to support the formulations of Henri and Michaelis-Menten and the idea of an enzyme-substrate complex, although the kinetics would still be consistent with adsorption catalysis. Direct evidence for the participation of the enzyme in the catalyzed reaction came from a number of approaches. From the 1930s analysis of the mode of inhibition of thiol enzymes—especially glyceraldehyde-phosphate dehydrogenase—by iodoacetate and heavy metals established that cysteinyl groups within the enzyme were essential for its catalytic function. The mechanism by which the SH group participated in the reaction was finally shown when sufficient quantities of purified G-3-PDH became available (Chapter 4).

A quite different approach came from Chance and others using heme enzymes (1947). Purified horseradish peroxidase has a characteristic absorption spectrum which was visibly altered in the presence of hydrogen peroxide. When an appropriate substrate was added it was oxidized by the hydrogen peroxide and the spectrum reverted to that of the original state of the enzyme. Similar studies were performed with catalase, showing that prosthetic groups in enzymes underwent reversible changes in the course of their reactions.

The development of nerve gases in World War II, especially di-isopropylphosphofluoridate, (DIPF), promoted urgent investigations of their bases of action. Adrian et al. and Mackworth showed esterases, particularly acetylcholinesterase, were strongly inhibited (1941). The range of hydrolases sensitive to DIPF was extensive and included chymotrypsin and trypsin, both of which were available in purified crystalline form. $DI^{32}PF$ was used to inhibit either enzyme, after which the proteins, now carrying ^{32}P, were repurified. They could be then sequenced and the site determined where the unique serine combined to the inhibitor (Jansen and Balls, 1949-1952). In this way the very large family of serine hydrolases were identified (For discussion of the early uses of DIPF, see Dixon and Webb, *The Enzymes*, 1959.)

The role of the serine residue in hydrolysis was further examined using pseudo-substrates, e.g. *p*-nitrophenylacetate—substrates which were only very slowly utilized by the enzyme. The p-nitrophenyl group was slowly released and the acyl group became attached to the same serine in hydrolases which had been detected by DIPF (Kilby and Youatt, 1954). Mechanisms for peptide and ester hydrolysis were therefore proposed in which the acyl group became transiently and co-valently bound to serines in catalytic sites (see Hartley et al. 1969).

From the realization of the amphoteric nature of proteins and the importance of $[H^+]$ in many of their functions, and as increasing numbers of enzymes were identified, it became commonplace to report their optimum pH. Corrections for non-enzymic hydrolysis of the substrates were usually made but, as was critically reviewed by Dixon and Webb (1959), from many of the results it was not possible to distinguish effects of pH on the protein in general from those on the region actually involved in interaction with the substrate. Nevertheless, many intracellular enzymes were found to have optimum pHs between 6 and 7.5. Examination of the pKs of different amino acid groups showed the pK for the imidazole group of histidine to be between 5.6 and 7.0. However, Cohn and Edsall (1943) and many others noted that the pKs of amino acids in proteins could be altered by interactions of their acid groups with neighboring residues.

Increased understanding of reaction mechanisms in the 1940s and 1950s pinpointed general acid or base catalysis as likely to be of importance in many hydrolytic reactions. The imidazole nucleus in histidine was the obvious center in proteins to donate or accept protons at physiological pH. The involvement of histidine was shown by photochemical oxidation in the presence of methylene blue (Weil and Buchert, 1951) which destroyed histidine and tryptophan and inactivated chymotrypsin and trypsin.

More specific evidence came from affinity labeling with molecules which could react with specific amino acid group sat or adjacent to the substrate site. These labels were substrate analogues and competitive inhibitors. Substituted aryl alkyl ketones were used. *N-p*-toluene-sulphonyl-L-phenylalanine chloromethyl ketone (TPCK) blocked the activity of chymotrypsin. Subsequent sequence analysis identified histidine 57 as its site of binding (see Hess, 1971, p 213, *The Enzymes*, 3rd ed.). Trypsin, with its preference for basic rather than aromatic residues adjacent to the peptide bond, was not blocked by TPCK but was susceptible to *N-p*-toluenesulphonyl-L-lysine chloromethyl ketone (TLCK) (Keil, ibid, p249).

A completely different and very dramatic identification of an enzyme active site was provided by David Phillips and his colleagues at the Royal Institution, London (1965-1967) for lysozyme. Lysozyme is an enzyme which hydrolyzes complex polysaccharide chains found in some bacterial cell walls. It is therefore potentially bacteriolytic. It was first studied by Laschtschenko in 1909 but became famous from the work of Fleming in the 1930s on the enzyme in tears which was feebly bacteriostatic. Hen egg-white lysozyme is competitively inhibited by N-acetylglucosamine trimer, GlcNAc-GlcNAc-GlcNAc. Phillips et al. crystallized the inhibited lysozyme. They could recognize the diffraction patterns from the bound GlcNAc trimer and thus localize the active site of the enzyme and determine the shape of the molecule and the position of its 129 amino acid residues.

Since 1960 our knowledge of protein structure and function has vastly increased. When the pathways were being elucidated, the enzymes involved were effectively "black boxes." Their catalytic activities were indubitable but their nature, mechanism of operation, and regulation were unknown, and in a sense irrelevant to the elucidation of the pathways. The discovery in the 1950s of the role amino acid residues played in enzyme activities was a vital stage in the reappraisal of protein structure. The proposal by Koshland (1958) that the position of the residues altered as the enzyme interacted with its substrate ("Induced Fit"), heralded a shift to a more dynamic view of proteins. This was to be further accelerated as the concept of allosterism (Monod, Changeux, and Jacob, 1963) received experimental support—for example, from studies on hemoglobin. Activation or inhibition of protein function by covalent modification, especially phosphorylation, was detected in the late 1950s. It has now emerged as the most obvious and universal mechanism for the regulation of protein function.

The study of proteins offers a useful illustration of the dependence of biochemistry on technical innovations. Improved methods for isolating proteins, easier ways to demonstrate their homogeneity, and the evolution of the Sanger procedure for sequencing, with its partial automation, greatly increased the number of proteins whose primary structure was known. Since 1960 technical advances have continued. In the protein field nuclear magnetic resonance (NMR) is being used to determine tertiary structure (molecular weight limit still ca. 20 kDa) and show the conformational flexibility of protein-ligand interactions. Probably the most far-reaching innovation has been the application of DNA sequencing to determine, from their cDNAs, the primary structure of proteins which may not yet have been isolated by classical procedures. Sophisticated computer programs enable data bases to be searched to identify commonly occurring domains within proteins which may have closely related conformations and functions.

Our knowledge of the diversity of protein functions has also expanded dramatically. Receptor molecules and the strategies adopted for communication across the cell membrane are increasingly well understood. Many proteins possess multiple surfaces for specific interactions with other macromolecules which allow molecular complexes and organelles to be assembled. More and more proteins are known to interact with and regulate expression from the eukaryotic genome. Enzymes involved in intermediary metabolism comprize only a small proportion of cellular proteins. To establish the links between the molecular biology of the genome and the classical metabolism of the cytoplasm is a task of the future.

REFERENCES

Bates, R.G. (1964). Determination of pH. Wiley & Son, New York.

Bergmann, M. & Fruton, J.S. (1941). The specificity of proteinases. Advances in Enzymol. 1, 63-98.

Cohn,E.J. & Edsall, J. (1943). Proteins, Amino Acids & Peptides. Reinhold Publishing, New York

Chibnall, A.C. (1966). The road to Cambridge. Annu. Rev. Biochem. 35, 1-22.

Corner, G.W. (1965). A History of the Rockefeller Institute, 1901-1953. The Rockefeller Institute Press, New York.

Davidson, J.N. (1950). The Biochemistry of the Nucleic Acids. Methuen, London.

Dixon, M. & Webb, E.C. (1958). The Enzymes. Longmans, Green & Co., London.

Ferdinand, W.H. (1976). The Enzyme Molecule. Wiley & Sons.

Fermi, G. & Perutz, M.F. (1981). Atlas of Molecular Structures in Biology. 2, Hemoglobin and Myoglobin. Clarendon press, Oxford.

Fersht, A. (1977). Enzyme Structure and Mechanism. W.H. Freeman, New York.

Haldane, J.B.S. (1930). The Enzymes. Longmans, Green London.

Jossang, P. (1992). Origins of chromatography. Nature, London 356, 100

Leggat Baily, J. (1962). Techniques in Protein Chemistry. Elsevier, Amsterdam.

Martin, A.J.P. & Synge, R.L.M. (1945). Analytical chemistry of proteins. Adv. Protein Chem. 2, 1-83.

Moore, S. & Stein, W.H. (1952). Ion-exchange chromatography. Annu. Rev. Biochem. 21, 521-546.

Northrop, J.H., Kunitz, M., & Herriott, R.M. (1939). Crystalline Enzymes. Columbia University Press, New York.

Sanger, F. (1952). Arrangements of amino acids in proteins. Adv. in Protein Chem. 7, 1-67.

Segal, H.L. (1959). The development of enzyme kinetics. In: The Enzymes (Boyer, P.D., Lardy, H., & Myrbach, K., Eds.), Vol. I, 2nd ed., pp. 1-48 Academic Press, New York.

Smithies, O. (1962). Starch gel electrophoresis. Archiv. Biochem. Biophys. Suppl. 1, 125-131.

Svedberg, T. & Pederson, K.O. (1940). The Ultracentrifuge. Oxford University Press.
Tiselius, A. (1947). Adsorption analysis of amino-acids. Adv. Protein Chem. 3, 67-93.
Young, E.G. (1963). Occurrence, Classification, Preparation and Analysis of Proteins. In Comprehensive Biochemistry (Florkin, M. & Stotz, E.H., Eds.), Vol. 7, pp. 1-55. Elsevier, Amsterdam.

Appendix 1

CHRONOLOGICAL SUMMARY OF MAIN EVENTS UP TO CA.1960

Pre-1900

Early 16th C.	*Leonardo da Vinci.* Animals unable to survive in an atmosphere that did not support combustion.
1679	*van Leeuwenhoek.* Lenses used to examine bacteria.
1780	*Laviosier and Laplace.* Amount oxygen utilized by guinea pigs directly proportional to their energy (heat) production
1828	*Wohler.* Synthesis of urea from ammonium cyanate.
1833-1838	Microscopic examination of plant and animal cells by *Dutrochet, Schlieden, and Schwann* led to acceptance of "the cellular origin of all tissues."
1837	*Berzelius* advanced the concept of catalysis.
1838	*Berzelius* coined the word "protein."
1842	*Liebig.* Carbohydrate, fat and protein oxidized in the body to release defined amounts of energy.
1857	*Pasteur.* Fermentation due to the presence of yeast cells.
1857-1866	*Voit.* Calorimetric studies to determine energy production by animals and man.
1861	*Graham* proposed proteins were colloids.
1869	Cell nuclei isolated by *Miescher.*
1878	Term "enzyme" coined by *Kuhne.*
1886	Cytochromes first described by *MacMunn.*

1849-1900	Endocrine secretions identified. The effects of gland extirpation corrected by grafts or extracts.
1897	*Edouard Buchner.* Intact yeast cells not required for fermentation, but only the zymases they contain.
1881-1905	*Lunin. Pekelharing.* Animals cannot survive on diets consisting solely of protein, carbohydrate, and fat.
1890-1897	*Eijkman.* Beri-beri induced in fowl by feeding polished rice, and corrected by an alcoholic extract from rice polishings.

1900-1960

1901	*Wroblewski* and
1908	*Harden and Young*. Inorganic phosphate essential for glycolysis.
1904	*Knoop.* ß oxidation of fatty acids.
1905	*Emil Fischer, Hofmeister.* Peptide structure of proteins proposed.
1906	*F.G.Hopkins.* Need for "accessory food factors" in diet.
1908	*Garrod.* "Inborn Errors of Metabolism."
1909	Concept of pH introduced by *Sorensen.*
1910	*Tsvett* used chromatography to separate chlorophylls.
1912-1948	Vitamins isolated. B vitamins shown to be parts of co-enzymes.
1913	*Michaelis and Menten.* Kinetics of enzyme action.
1921	*Banting and Best.* Highly purified insulin used to treat diabetics.
1923	*Hevesy* first used a radioisotope as a tracer.
1924	*Warburg.* Importance of iron pigments in oxidation.
1925	*Keilin* rediscovered cytochromes.
1926	*Sumner* crystallized urease.
1927	The *Eggletons* and *Fiske and SubbaRow* isolated phospho-creatine.
1928	*Lohman* discovered ATP in muscles.
1932	*Krebs and Henseleit.* The urea cycle.
1934	*Svedberg* began studies with the ultracentrifuge.

1933-1934	*Peters* showed vitamin B_1 required for the oxidation of pyruvate *in vitro*.
1935-1936	Dicarboxylic acid cycle discovered by *Szent-Gyorgi*.
1937	*Krebs and Johnson*. The tricarboxylic (citric) acid cycle.
1937	*Braunstein and Kritzman* discovered transamination.
1930-1939	Early demonstrations that phosphate esterification was linked to oxygen uptake.
1939-1941	*Schoenheimer and Rittenberg*. "The Dynamic State of Body Constituents."
1939-1941	*Hevesy* demonstrated the permeability of cell membranes to Na^+ and K^+.
1939-1941	*Engelgardt and Ljubimova* found the structural protein myosin to be an ATPase.
1941	*Lipmann*. The "High Energy Phosphate" concept.
1941-1943	*Banga and Szent-Gyorgi*. Actomyosin threads contracted in the presence of ATP.
1941	*D.D. Woods*. Prontosil bacteriostatic because its active principle inhibited the utilization of *p*-aminobenzoic acid.
1943-1944	*Martin and Synge*. Paper chromatography introduced.
from 1948	Methods for the isolation of mitochondria became available.
1948-1950	*Lehninger and Kennedy* demonstrated oxidative phosphorylation in mitochondria.
1950	*Leloir* showed uridine diphosphate sugars needed for polysaccharide syntheses.
1946-1953	*Calvin* established the pathway for carbon dioxide fixation in higher plants.
1951	*Moore and Stein* introduced quantitative chromatographic amino acid analysis for proteins.
1952-1953	*Lynen*. The fatty acid (oxidation) spiral.
1953	*Watson and Crick*. The DNA double-helix.
1953	*Sanger*. Primary structure of insulin established.
1953-1954	*A. Huxley and Niedergerke, H.E.Huxley and Hanson*. The sliding filament theory of muscle contraction.

1953-1955	Electron micrographs of cell ultrastructure become available.
1955	*Wakil.* Fatty acid synthesis dependent on HCO_3^-.
1955	*Chance.* Double-beam spectophotometer introduced.
1956	*Sutherland.* Discovery of cyclic AMP.
1957	*Skou* identified the "sodium pump" as a Na^+/K^+ dependent ATPase in the cell membrane.
1958	*Koshland.* "Induced fit" suggested for protein-ligand interactions.

Post-1960

1961-1966	*Mitchell.* The chemiosmotic theory of oxidative phosphorylation.
1963	*Monod, Changeux and Jacob* introduced the concept of allosterism.
1972	*Singer and Nicholson's* fluid-mosaic theory of membrane structure.
1972	*Blundell, Dodson, Hodgkin and Mercola* reported the three-dimensional structure of insulin.

Appendix 2

PRINCIPAL METABOLIC PATHWAYS

1. Glycolysis
2. Glycogen Metabolism
3. The Tricarboxylic (Citric) Acid Cycle
4. The Urea Cycle
5. Fatty Acid Oxidation Spiral
6. Animal Fatty Acid Synthesis

Glycolysis

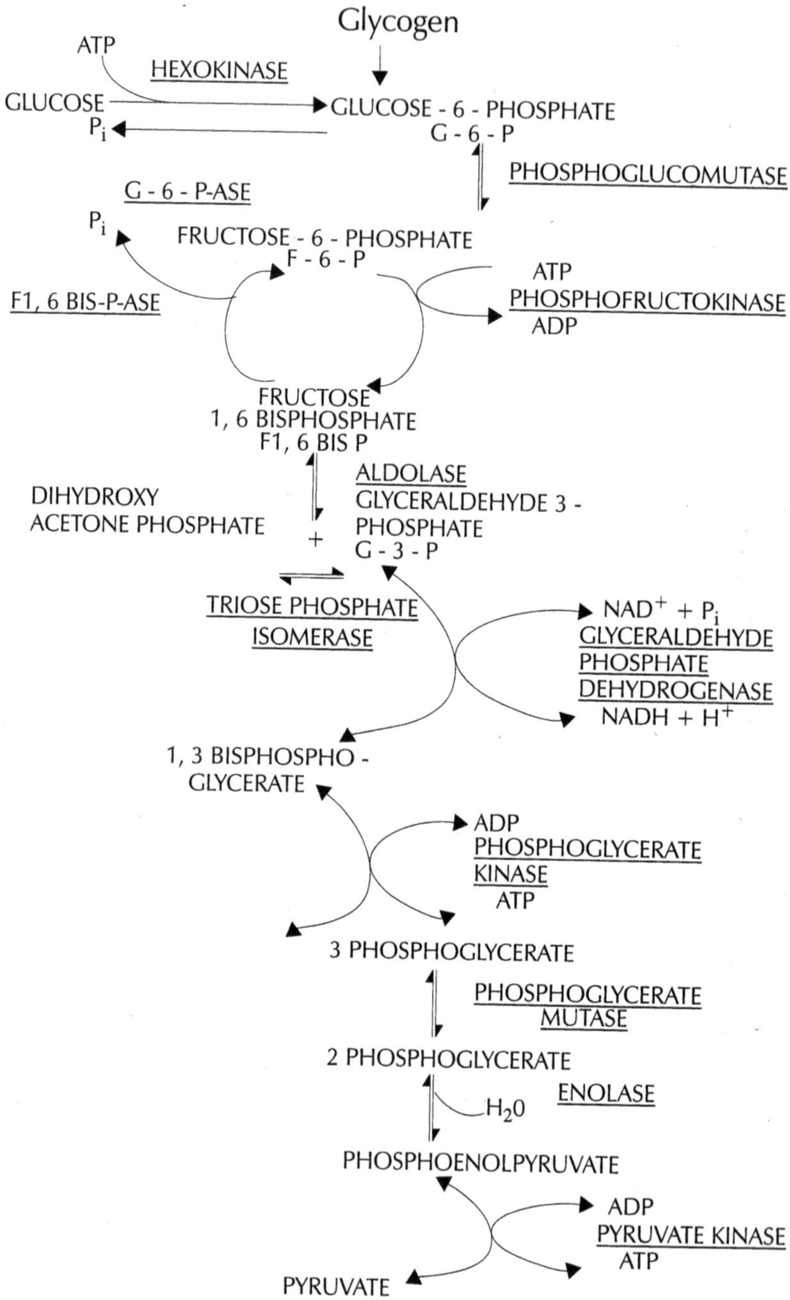

Glycogen
ATP
HEXOKINASE
GLUCOSE
GLUCOSE - 6 - PHOSPHATE
P_i
G - 6 - P
PHOSPHOGLUCOMUTASE
G - 6 - P-ASE
P_i
FRUCTOSE - 6 - PHOSPHATE
F - 6 - P
ATP
PHOSPHOFRUCTOKINASE
ADP
F1, 6 BIS-P-ASE
FRUCTOSE
1, 6 BISPHOSPHATE
F1, 6 BIS P
ALDOLASE
DIHYDROXY
ACETONE PHOSPHATE
+
GLYCERALDEHYDE 3 -
PHOSPHATE
G - 3 - P
TRIOSE PHOSPHATE
ISOMERASE
$NAD^+ + P_i$
GLYCERALDEHYDE
PHOSPHATE
DEHYDROGENASE
$NADH + H^+$
1, 3 BISPHOSPHO -
GLYCERATE
ADP
PHOSPHOGLYCERATE
KINASE
ATP
3 PHOSPHOGLYCERATE
PHOSPHOGLYCERATE
MUTASE
2 PHOSPHOGLYCERATE
H_20
ENOLASE
PHOSPHOENOLPYRUVATE
ADP
PYRUVATE KINASE
ATP
PYRUVATE

Glycogen Metabolism

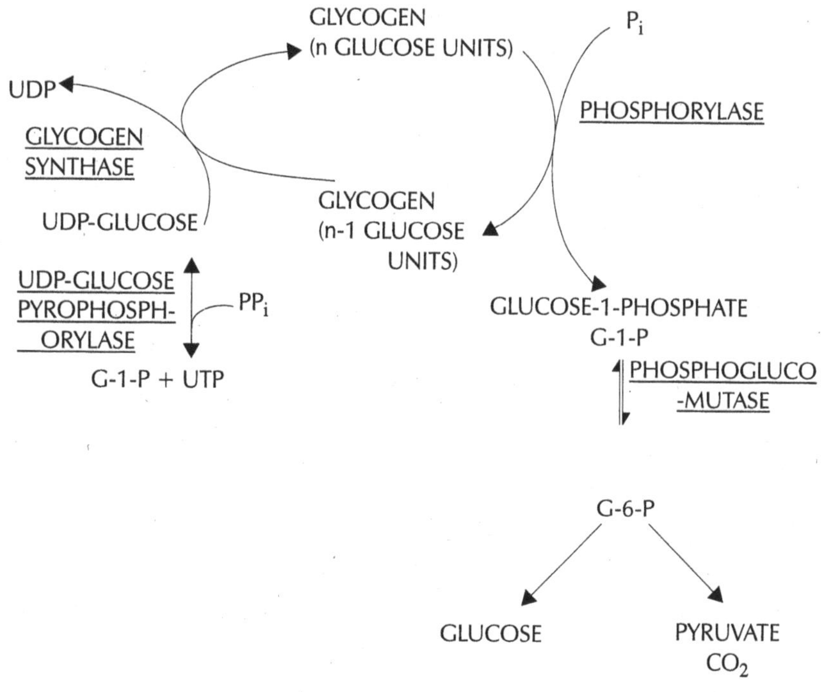

The Tricarboxylic (Citric) Acid Cycle

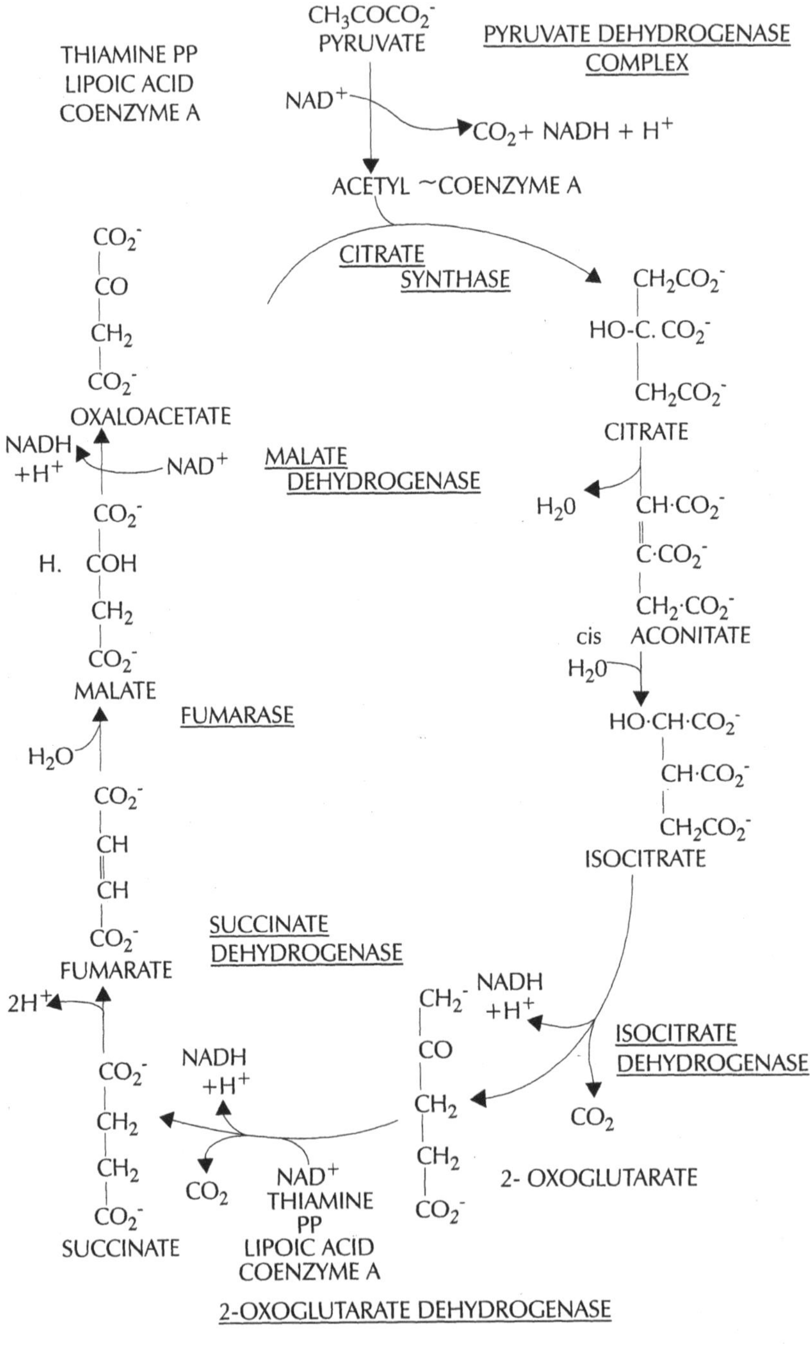

Urea Cycle

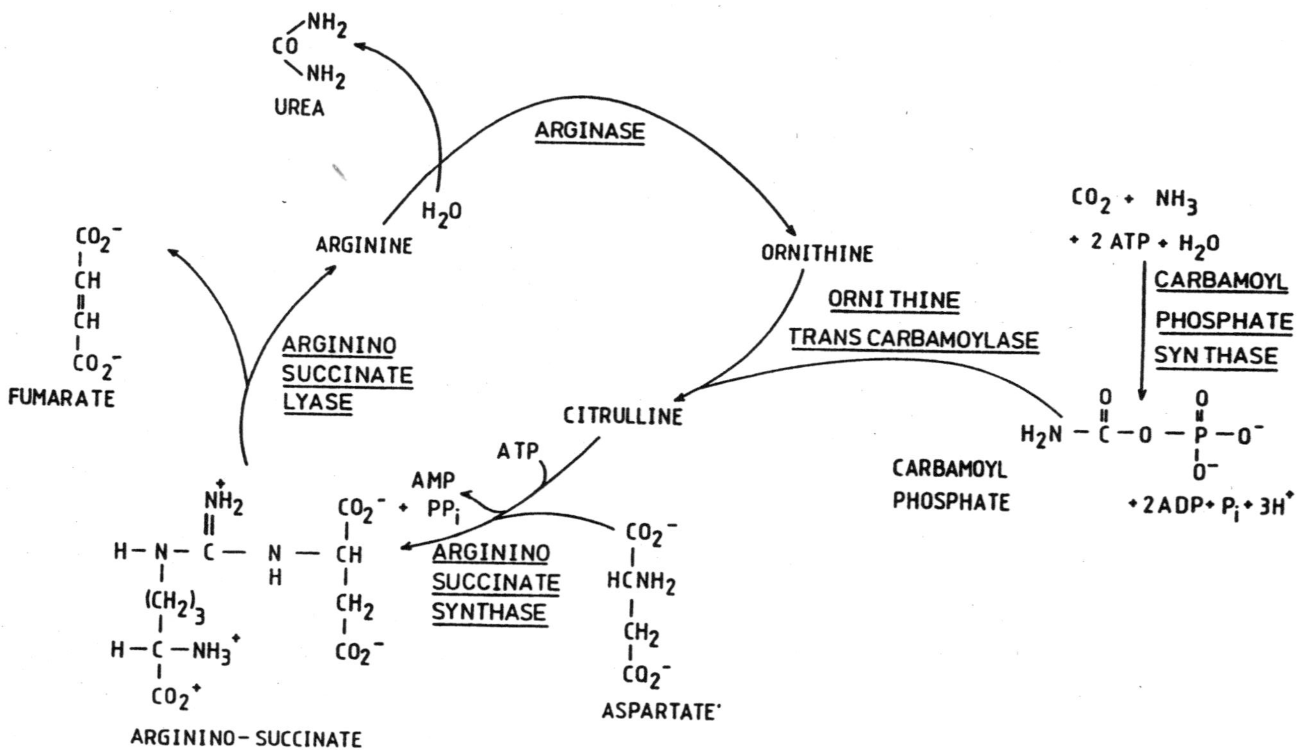
UREA
ARGINASE
H_2O
ARGININE
ORNITHINE
CO_2 + NH_3
+ 2 ATP + H_2O
CARBAMOYL
PHOSPHATE
SYN THASE
ORNI THINE
TRANS CARBAMOYLASE
FUMARATE
ARGININO
SUCCINATE
LYASE
CITRULLINE
ATP
AMP
PP_i
CARBAMOYL
PHOSPHATE
+ 2 ADP + P_i + $3H^+$
ARGININO
SUCCINATE
SYNTHASE
ASPARTATE
ARGININO-SUCCINATE

Fatty Acid Oxidation Spiral

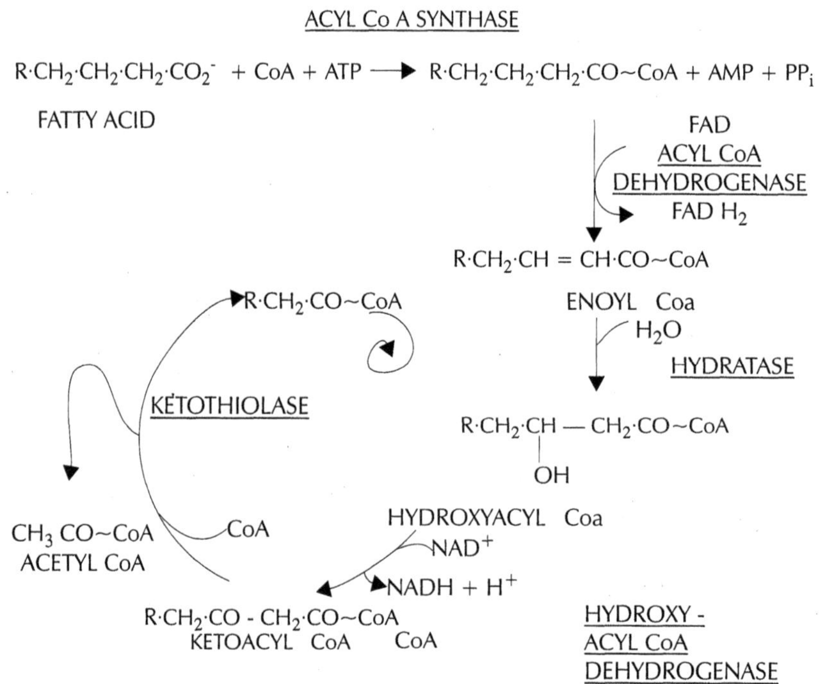

Animal Fatty Acid Synthesis

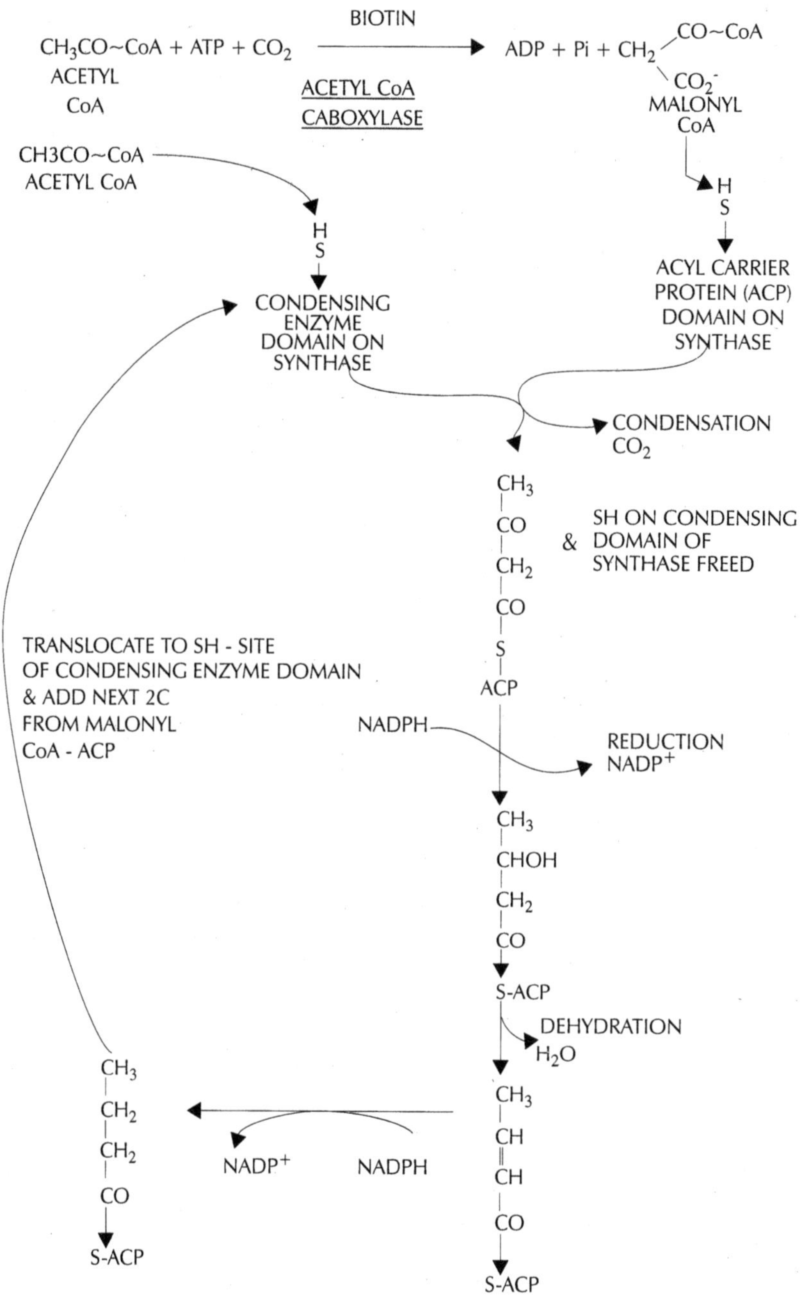

AUTHOR INDEX

SUBJECT INDEX

www.ingramcontent.com/pod-product-compliance
Ingram Content Group UK Ltd.
Pitfield, Milton Keynes, MK11 3LW, UK
UKHW020810290926
15470UKWH00003B/1058

9 781559 389600